When?

Experiments for the Young Scientist

Robert W. Wood
Illustrated by Steve Hoeft

TAB Books
Division of McGraw-Hill, Inc.
New York   San Francisco   Washington, D.C.   Auckland   Bogotá
Caracas   Lisbon   London   Madrid   Mexico City   Milan
Montreal   New Delhi   San Juan   Singapore
Sydney   Tokyo   Toronto

© 1995 by Robert W. Wood.
Published by TAB Books.
TAB Books is a division of McGraw-Hill, Inc.

pbk 1 2 3 4 5 6 7 8 9 DOH/DOH 9 9 8 7 6 5 4

Library of Congress Cataloging-in-Publication Data

Wood, Robert W., 1933-
  When? : experiments for the young scientist / by Robert W. Wood;
illustrations by Steve Hoeft.
     p. cm.
   Includes index.
   ISBN 0-07-051640-5
   1. Science—Experiments—Juvenile literature. 2. Scientific
recreations—Juvenile literature. [1. Science—Experiments.
2. Experiments.] I. Hoeft, Steve, ill. II. Title.
Q164.W68 1994
507.8—dc20   94-10825
                                                CIP
                                                AC

Acquisitions Editor: Kimberly Tabor
Editorial team: Steven Bolt, Editor
              Susan W. Kagey, Managing Editor
              Joanne M. Slike, Executive Editor
              Joann Woy, Indexer
Production team: Katherine G. Brown, Director
              Susan E. Hansford, Coding
              Rose McFarland, Desktop Operator
              Ruth Gunnett, Computer Artist
              Nancy K. Mickley, Proofreader
Design team: Jaclyn J. Boone, Designer   0516405
              Brian Allison, Associate Designer   KIDS

Contents

THE YOUNG METEOROLOGIST

THE YOUNG BIOLOGIST

THE YOUNG PHYSICIST

Introduction

Science began thousands of years ago, before anyone learned to write. We'll never know who discovered fire, invented the wheel, or tried to explain what the Sun is and where it went during the night. In general, mathematics and engineering, followed by the physical and biological sciences, were probably the first of the sciences to be developed.

In ancient times, things had to be learned mostly by trying over and over again until the project was successful. But today you can go to school to easily learn what were difficult lessons in the past. After a problem is solved, you can use several steps in the scientific method to explain a problem and its solution.

1. Ask a question or state a problem. For example, is there a connection between the time of day and the formation of rainbows?
2. Form the hypothesis or possible explanation. In the example, you know that rainbows are formed when sunlight strikes water droplets in the sky, and that you only see rainbows in mornings and afternoons. This can lead you to believe that there is a connection between the angle of the Sun's rays and the water droplets.
3. Experiment. In the preceding example, in mid-morning or afternoon, you could go outside and adjust the spray of a garden hose to a fine mist. Spray the mist in the air until you produce a rainbow. You will see that the rainbow only appears when sunlight is behind you.
4. Interpret data and form a conclusion. From the experiment, you will realize that sunlight is reflected and refracted back from water droplets to form a rainbow. The time of day has a definite connection to the formation of rainbows.

This is one of a series of books that asks *Why?*, *What?*, *When?*, *Who?*, and *Where?* science happens, and it discusses some of the people who made science happen. Each book is divided into six parts; each part relates to a different science. You should know, however, that sciences tend to overlap from one field to another. A knowledge of physics, for example, will be useful in the study of biology as well as astronomy.

Each part of this book consists of easy experiments that begin by asking a question, followed by the materials list, a step-by-step procedure and the results. At the end of each experiment you will find suggestions for further studies that will broaden the scope of the experiment. Also included are science trivia and oddities for your amusement.

Symbols used in this book

 adult supervision

 scissors

 sharp object

 safety goggles

 fire

 electricity

 stove

Part 1
The young engineer

Engineers are builders who apply their skills and scientific knowledge to use power and materials in the most effective ways for projects that benefit society. Engineers design and supervise the construction of dams, electricity-generating plants, highways, and skyscrapers, as well as airplanes, cars, and other forms of transportation. The space shuttle is an excellent example of what the combined talents of today's engineers can produce.

Engineers have developed new medical technology, such as *magnetic resonance imagery* (MRI), that is used to look inside our bodies without surgery. Through the science of engineering we have increased our comfort, safety, and overall lifestyle. Simply put, engineers use scientific knowledge to make the best use of energy and materials.

1
When was the first balloon flight?

materials ☆ ❏ Hair dryer
❏ A clear, plastic dry cleaner's garment bag

procedure ☆ 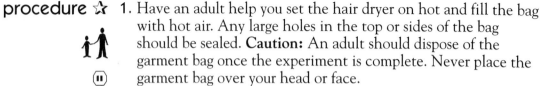 1. Have an adult help you set the hair dryer on hot and fill the bag with hot air. Any large holes in the top or sides of the bag should be sealed. **Caution:** An adult should dispose of the garment bag once the experiment is complete. Never place the garment bag over your head or face.

2. When the bag is fully inflated, release it. What happens?

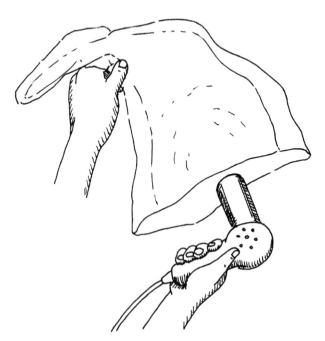

Fill the bag with hot air.

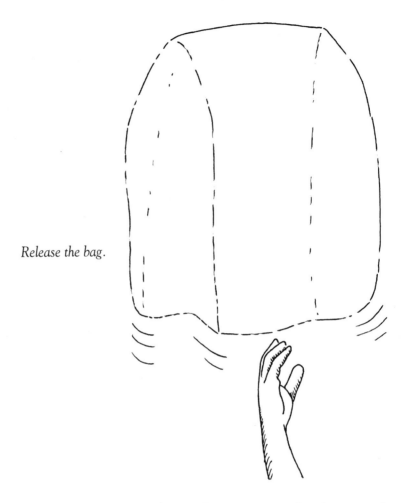

Release the bag.

results ☆ The bag slowly rises into the air. Because air molecules expand when heated, they occupy a much larger space than before. The bag is less influenced by gravity than the surrounding air. The heavier, cooler, surrounding air forces the warmer air upward.

Two French paper manufacturers, brothers Joseph and Etienne Montgolfier, noticed how smoke rises in the air. They thought that if they could collect enough smoke in a bag, the bag would go up. After some experimenting, they built a bag, made of taffeta and lined with paper, 35 feet (10.7 meters) in diameter and weighing about 300 pounds.

The Montgolfier balloon

On June 5, 1783, the Montgolfier brothers staged the first balloon exhibition. They built a fire of wool and straw under the opening of the bag to fill it with smoke. To the surprise of spectators, the bag rose to a height of about 6000 feet, drifted about a mile, then settled back to earth. The bag rose because air expands and becomes lighter when heated.

Joseph Montgolfier continued experimenting and built a balloon 86 feet (26.23 meters) tall and 48 feet (14.6 meters) across that had a wicker basket for passengers. On November 21, 1783, the hot air from a fire of straw and wool inflated the balloon, and it lifted two men, Pilatre de Rozier and Marquis d'Arlandes, about 300 feet (92 meters) over Paris.

Today, a balloon can be filled with hot air from the flames of a propane burner.

Try your hot air balloon on a cool, calm day. Does it rise faster? Tie the open end of your balloon closed and see if it travels farther. Helium is used to inflate balloons at a circus. Is helium lighter than air? Compare the flight of a hot air balloon with a helium-filled balloon. If you blow up a balloon using your lungs, does the balloon settle to the ground? Is compressed air heavier than the surrounding air?

did you ☆
know?

❏ That the first balloon flight was thought to be so dangerous that the original plan called for two criminals to be the passengers.

❏ That George Washington witnessed the first successful balloon flight in America in Philadelphia on January 9, 1793. The balloonist was Jean-Pierre Blanchard, from France.

❏ That a balloon flight to study meteorology was made by James Glaisher and Henry Coxwell in England in September 1862. After reaching an altitude of about 37,000 feet (11,285 meters), Glaisher and Coxwell became unconscious from the lack of oxygen, but they later revived when the balloon started to sink.

2
When were airships popular?

materials ☆
- ❒ Medium-sized helium balloon (purchased from a retail shop)
- ❒ Length of ribbon or sewing thread
- ❒ Key or nail for weight
- ❒ Paper clips for ballast
- ❒ Transparent tape

procedure ☆

1. Your balloon will probably come with a length of narrow ribbon, so you might ask for a ribbon about 6 feet (1.8 meters) long. If you don't have enough ribbon, use thread. Form a small loop from the bottom of the balloon to the top and fasten it in place with a short piece of tape. About 5 feet (1.5 meters) of ribbon will be left over for a tether.

2. Attach paper clips in a chain to the loop until the balloon balances level and floats in the air without rising or falling. Tie the key or nail to the free end of the ribbon for the anchor. Except for the lack of an engine, you now have an airship similar to those of earlier years. Place your airship in a room with a slight draft. What happens?

results ☆ The airship slowly rises and then settles back toward the floor, and then rises again. The movement is determined by the amount of draft in the room. As the airship settles toward the floor, the weight of the tether is removed from the load of the ship and it starts to rise again. If the airship settles down enough for the paper clips to touch the floor, the weight of the clips is removed from the load and the airship rises again. Gradually, some of the helium escapes, and you have to remove a paper clip to keep the balance. If one paper clip is too much weight, use wire ties from a garbage bag or a bread bag.

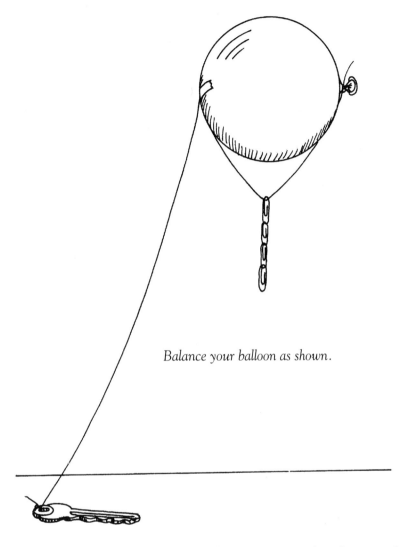

Balance your balloon as shown.

Airships were developed in two different types: rigid and nonrigid. *Rigid airships* have metal frames to hold their shapes. *Nonrigid airships* maintain their shapes by the pressure of the helium gas inside. The loss of the *Hindenburg* and other disasters put an end to the production of rigid airships. The nonrigid airship, the only type used today, is usually called a *blimp*. Blimps were used successfully in World War II. Today, we often see Goodyear and other company-sponsored blimps at sporting events serving as TV broadcasting and advertising platforms. Although the Goodyear company no longer builds blimps, it still operates three throughout the United States.

further ☆ Try flying your airship over a lighted lamp. Does the heat from the
studies lamp create updrafts, causing the ship to rise? Compare flying from
a calm room to a drafty one. Are winds a problem for airships?
Once the airship is filled with helium, you can see that it takes very
little energy to lift a load. Could airships be used as sky cranes?

did you ☆ ❒ That balloons are used to remove logs from some rough
know? mountainous areas in logging operations?

❒ That Henri Giffard, a French engineer, was the first man to
pilot a power-driven airship. Powered by a 3-horsepower steam
engine, Giffard's airship was 143 feet (43.7 meters) long and 39
feet (11.9 meters) in diameter. Giffard flew over Paris in 1852.

❒ That on October 19, 1901, Alberto Santos-Dumont, a Brazilian
living in Paris, won a 100,000-franc prize for navigating his
airship from St. Cloud, around the Eiffel Tower, and back. His
airship was powered by a 12-horsepower motorcycle engine.

❒ That early European airships were filled with explosive
hydrogen gas, and that this gas caused the German airship
Hindenburg to explode with a loss of 35 lives at Lakehurst, New
Jersey, in 1937. Helium gas, which is almost as light as
hydrogen and is noncombustible, was used in American
airships.

3
When was the
first airplane flown?

materials ☆ ❏ Sheet of 8½-x-11-inch paper
❏ Paper clip
❏ Scissors

procedure ☆ 1. Fold the paper in half lengthwise, and then spread it out flat
again. Now fold the corners of one end in towards the center;
make sure the creases are sharp.

2. Fold the outside corners in so that they meet in the center once
more. Fold the paper back in half along the first fold you made.
Make the wings by folding each half of the paper down the
same distance.

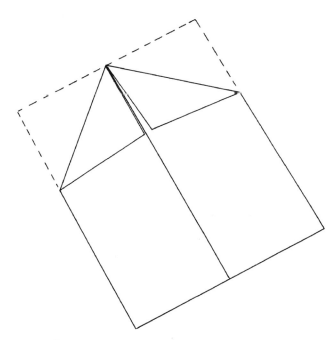

Fold the corners down.

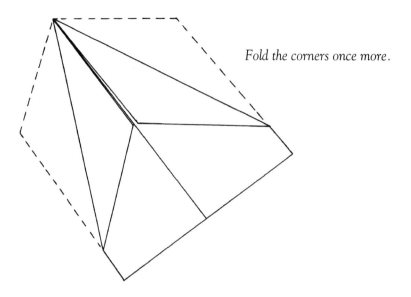

Fold the corners once more.

3. Remembering to always point the scissors away from yourself, cut two small slits in the back of each wing for up and down control. Cut two more slits in the back of the body (fuselage) for a rudder. Attach a paper clip about halfway down the fuselage for balance. Take your airplane outside and toss it into the air. How does it fly? Does it go straight ahead, up, or down?

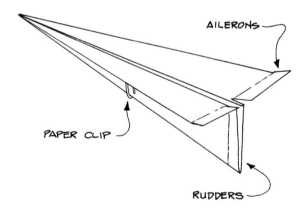

Attach a paper clip for weight and balance.

further ☆ Move the paper clip back or forward along the fuselage to change
studies the balance. Does this change the flight pattern? Experiment by adding more paper clips. Next, remove all of the paper clips and

try it again. What effect does the weight have on the gliding distances? Bend the flaps on the wings up or down to make the airplane climb or dive. Experiment to find the best weights and flap positions for the longest flights. These flaps are called *ailerons* when they are on the wings of conventional aircraft and *elevators* when they are on the tail. On your airplane, a delta wing, they are combined (ailerons and elevators) and are called *elevons*. Bend the *rudder* to make the airplane turn right or left.

did you ☆ **know?**

❏ That in 1891, Otto Lilienthal built a glider of willow rods and covered it with cotton fabric. It had over 100 square feet (9.3 square meters) of lifting surface and weighed about 40 pounds. He was supported by a padded opening in the frame. Much like the way today's hang gliders are controlled, Lilienthal was able to control his glider by shifting his weight.

❏ That on December 17, 1903, at Kitty Hawk, North Carolina, Orville Wright made a successful flight in a machine under its own power. The flight lasted only 12 seconds and covered a little over 120 feet (36.6 meters). His brother, Wilbur Wright, later made a flight lasting almost a minute and covering 852 feet (260 meters). Which brother flew first was determined by the toss of a coin.

❏ That in 1909 the U.S. Army Signal Corps purchased the first military airplane in the world from the Wright brothers. The specifications called for the airplane to have a speed of at least 40 miles (64.4 km) per hour, make a cross-country flight of 10 miles (16 km) and stay in the air for one hour. Another requirement was that the craft be capable of carrying a passenger. The price was $25,000 with a bonus of $5,000 for exceeding the speed by two miles (3.2 km) per hour.

4
When was the first parachute jump made?

materials ☆ ❏ Handkerchief
❏ Four strings, each 9 or 10 inches (250 mm) long
❏ Weight (medium-sized fishing sinker or two or three washers)

procedure ☆ 1. Tie one end of the strings to each corner of the handkerchief.
2. Suspend the handkerchief from its center and bring the four strings together so that their lengths are equal. Now thread the ends through the opening in the weight and fasten it with a knot. You now have a parachute.

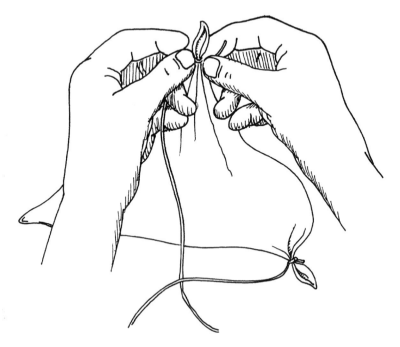

Tie strings to the corners.

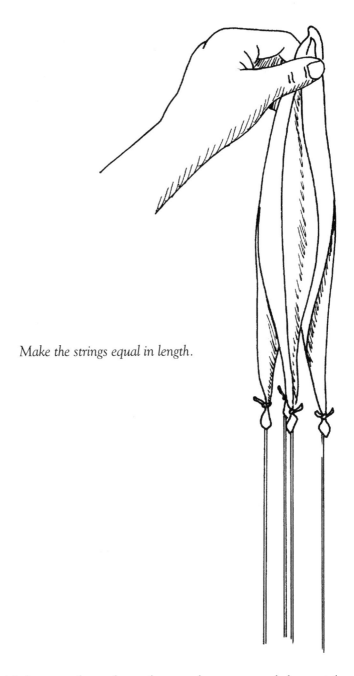

Make the strings equal in length.

3. Fold the parachute from the top down, toward the weight, wrapping the strings around the cloth into a small bundle. Toss it high into the air. Does your parachute work?

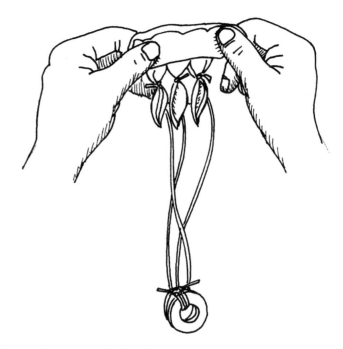

Fold the parachute in a tight bundle.

results ☆ The parachute bundle travels up to a point then starts to fall. The weight pulls the parachute open and allows the chute to slow the descent. When the parachute is tossed as a small bundle, the resistance of the air is little. When the chute opens and catches the air, the resistance is greatly increased, and thereby slows the fall.

further studies ☆ Try using smaller, then larger, weights. What happens to the rate of descent? Ask an adult for permission to use a vacuum cleaner. Connect the hose of a tank-type vacuum cleaner to the exhaust and support your parachute on the column of air. Try larger and lighter materials (clear plastic wrap, for example) for the parachute canopy. Does this make your parachute more efficient?

did you know? ☆ ❏ That Leonardo da Vinci designed a parachute he called a "tent roof" in 1495.
❏ That the first successful parachute jump was made from a tower in 1783 by the French physicist Sebastion Lenormand.
❏ That over Paris on October 22, 1797, Andre Jacques Garnerin was the first man to drop from a balloon in a parachute.

❐ That the first man to parachute from an airplane was Captain
Albert Berry. He jumped on March 1, 1912, from a Benoist
Pusher, flying 1500 feet (475.5 meters) over Missouri, at a
speed of 50 miles (80.45 km) per hour. The airplane was piloted
by Antony Jannus.

5
When was the first jet airplane flown?

materials ☆
- ❏ Two chairs
- ❏ String, about 15 feet (4.6 meters) long
- ❏ Drinking straw
- ❏ Balloon
- ❏ Transparent tape

procedure ☆ 1. Place the chairs about 15 feet (4.6 meters) apart. Feed one end of the string through the straw and fasten each end of the string to the tops of the chairs.

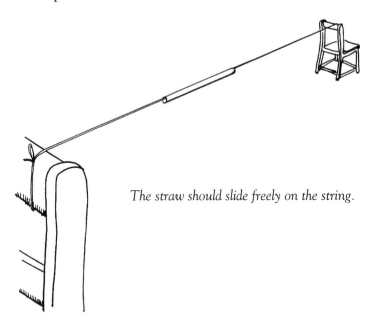

The straw should slide freely on the string.

2. Blow up the balloon and tie off the opening with a string tied in a slip knot. The knot should be easily untied, but make several wraps with the string before tying the knot.

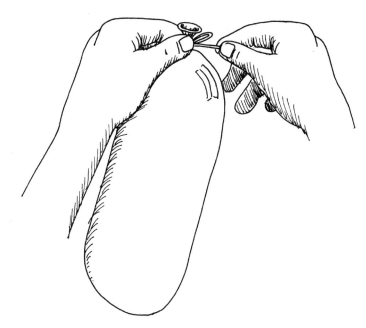

Tie the string in a slip knot.

3. Tape the balloon to the straw in a couple of places and slide it near the chair on the side of the opening of the balloon. Release the knot. What happens?

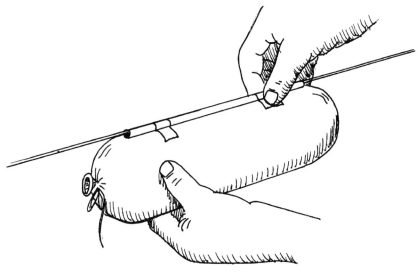

Fasten the balloon to the straw.

results ☆ When the air in the balloon is released, the balloon shoots down the string to the other chair. The reason the balloon moves is because every action has an equal and opposite reaction (Newton's third law). The air pressure in the balloon produces an exhaust from the opening. This exhaust from the back of the balloon pushes against the stationary air in the room to create a force called *thrust*. The thrust causes the balloon to move forward.

Jet aircraft fly by the same principle. Air is drawn in the front of the engine and compressed, which heats the air to a very high temperature. The air is further heated by injecting fuel. The hot gas, under high pressure, pushes out through a nozzle at the rear to provide the thrust.

further ☆ Place a garden hose in a clear area outside and turn the water on
studies full. Will the end of the hose be pushed about? Is this an example of thrust? Notice a jet flying at a high altitude. Can you see a vapor trail from the exhausts? Propellers lose their efficiency at about 300 miles (483 km) per hour, which limits the performance of propeller-driven aircraft. Is efficiency the main reason that jets can fly higher and faster than propeller-driven aircraft?

did you ☆ ❏ That as early as 1939, a jet-powered airplane was flown in
know? Germany.
❏ That the first jet airplane designed and built in the United States was the XP-59. It was flown for the first time on October 1, 1942, at Muroc, California, by Robert Stanley. It was rated over 400 miles (644 km) per hour with an altitude of over 40,000 feet (12,200 meters).
❏ That the first man to fly faster than the speed of sound (Mach 1.0 or 760 miles—1223 km—an hour at sea level) was Charles ("Chuck") Yeager on October 14, 1947. The aircraft, the rocket-powered Bell X-1, was dropped from an Air Force bomber at a high altitude over Edwards Air Force Base, Muroc, California. The speed was 967 miles (1556 km) per hour at an altitude of 70,140 feet (21,393 meters).
❏ That in the late 1950s and mid-1960s a research aircraft, the North American X-15, reached speeds of Mach 6.72 (4534 mph or 7297 km) and flew at altitudes of about 70 miles (113 km) above the earth.

6
When was the zipper first used?

materials ☆ ❑ Large zipper on a heavy jacket or luggage
❑ Magnifying glass

procedure ☆ 1. With the zipper opened, place the two sides close together and examine the edges with a magnifying glass.
2. Slowly draw the slide to close the zipper and watch how the teeth mesh together.
3. With the zipper closed, gently try to separate the two sides.

The zipper teeth mesh together.

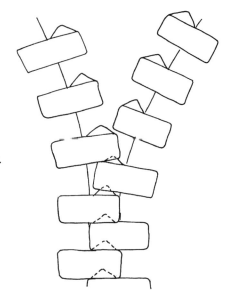

results ☆ Under the magnifying glass, you can see that the teeth on each side have hollows on the bottoms and points on the top. When you close the zipper, you can see how the teeth mesh together with the points on one side fitting into the hollows on the other side. The teeth fit so snug that the sides are firmly locked together.

further ☆ How many uses can you find for zippers around your home? Do
studies your examples include clothing, purses, luggage, boots, sleeping
bags, and back windows in convertibles? Is Velcro being used to
replace some zippers?

did you ☆ ❏ That Whitcomb L. Judson, of Chicago, received a patent on
know? August 29, 1893, for a zipper fastener. It had a series of hooks
and eyes that fastened together with a slider.
❏ That Lewis Walker, Judson's friend, received a patent on the
meshed-tooth type of slide fastener in 1913.
❏ That between 1913 and 1917, Gideon Sundback of Hoboken,
New Jersey, received patents improving "separable fasteners"
that were mass-produced during World War I.

7
When was the steam engine first put to use?

materials ☆
- ☐ Water
- ☐ Sauce pan
- ☐ Matching lid with insulated handle
- ☐ Stove
- ☐ Kitchen sink

procedure ☆

1. Ask an adult to help you with this procedure. Fill the pan with a couple inches (50 mm) of water.
2. Place the lid on the pan, grasp the handle, and slowly rotate the lid.
3. Place the pan on the stove and bring the water to a boil.
4. Being careful of the escaping cloud of vapor, slowly rotate the lid again. Be sure to use a hot pad or oven mitt to protect your hand.

The lid now turns easily.

5. While the water is still boiling, remove the lid and observe the bubbles. Replace the lid.
6. Remove the pan from the stove, hold it over the sink, and cover the pan with cool running water.
7. Gently try to lift the lid.

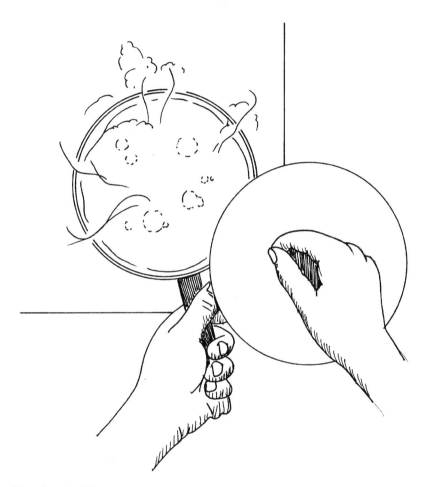

The clear bubbles are steam.

results ☆ Before the water is heated, the lid resists rotation. When the water boils, the lid rotates easily because it floats on a column of steam. With the lid removed, you can see clear bubbles of steam develop in the water. When the pan cools, the steam condenses inside the pan and creates a partial vacuum.

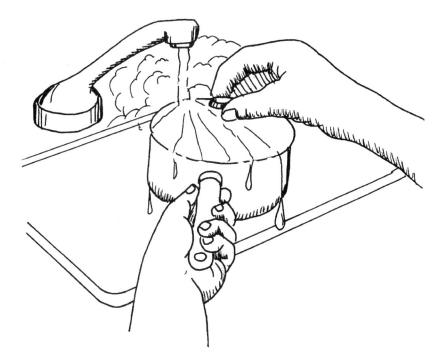

Cold water creates a small vacuum.

The first steam engines worked on the capability of steam to condense rather than its capability to expand. In 1698, Englishman Thomas Savery received the patent on the first practical steam engine. Savery's engine was designed as a pump to suck water from mines. It had no moving parts except for valves operated by hand. Steam was allowed to enter a sealed container. When the container was cooled by cold water, a vacuum was created. The power of the vacuum was used to suck water up a pipe lowered into the mine.

In 1712, Thomas Newcomen, an English blacksmith, invented an improved model of the steam-engine pump. Newcomen's pump used a piston fitted inside a cylinder connected to a large horizontal beam balanced in the middle. Steam was let into the cylinder, forcing the piston up and lowering the other end of the beam. Cold water then cooled the cylinder and the vacuum pulled the piston back down.

In 1769, Scottish engineer, James Watt took out a patent on a much more efficient steam engine that had a separate condenser and cylinder. Watt's design eliminated the need for alternately heating and cooling the cylinder.

James Watt

further studies ☆ Think about the different ways steam is used today. How many can you list? Consider the fuels we use now and the pollution in the air. Could steam be an alternate source of power? Remember, you still must heat the water somehow. Can you heat water using sunlight?

did you know? ☆
❐ That steam cannot be seen. The cloud of vapor you see coming from a teakettle is moisture that the cooler air has changed, or condensed, from the gas into tiny particles of water. The steam is in the clear space between the spout and the vapor.
❐ That James Watt was the first person to develop a practical way of using steam for heat. In 1784, he used steam contained in coils of pipe to heat his office.
❐ That steam turbines were invented in the late 1800s.

Part 2
The young astronomer

Astronomy is the study of the heavenly bodies and their movements. It is probably one of the oldest sciences. As early tribes became farmers, they needed to know the seasons for planting and harvest. This need led to the development of various types of calendars that required celestial observations. Later, calendars were made based on the movements of the Sun during the year. Even today, our clocks are regulated by the movement of the Sun.

Our solar system is made up of the Sun and the planets and other heavenly bodies that revolve around the Sun. Our solar system is part of a giant galaxy called the Milky Way. The Sun is a little over halfway from the center of an outer arm of the Milky Way, and revolves around the center of the mass of whirling stars at about 175 miles (282 km) per second.

The Earth and the rest of our solar system tag along with the Sun at this dazzling speed, but it still takes more than 200 million years for our solar system to complete just one revolution around the Milky Way. Our knowledge of the universe is constantly growing through advances made in astronomy. You can begin this exciting hobby with good eyesight and a clear night.

8
When was the telescope first used?

materials ☆
- ❏ Curved shaving or make-up mirror
- ❏ Table or stand
- ❏ Window
- ❏ Flat pocket mirror
- ❏ Moonlit night
- ❏ Magnifying glass

procedure ☆
1. Place the shaving mirror on a table or stand it in front of a window with the magnifying side facing the Moon.

Use a magnifying glass to look at the image in the flat mirror.

2. Position the flat mirror facing the shaving mirror so that you can see a reflection of the Moon in the flat mirror.
3. Use the magnifying glass to look at the image of the Moon in the flat mirror.

results ☆ Light traveling from the Moon strikes the large curved surface of the shaving mirror. The light is then reflected to the small flat mirror, where it is once more reflected through the magnifying glass. The large curved mirror is used to gather moonlight and concentrate it on the smaller flat mirror. The magnifying glass enlarges the image of the Moon. This type of telescope is called a *reflecting telescope* because it uses a large mirror to gather light from an image. The simplest type of telescope is called the *refracting telescope*. It uses only lenses.

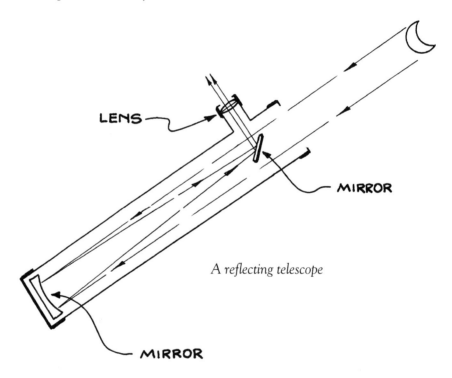

LENS

MIRROR

A reflecting telescope

MIRROR

The first telescope was invented in 1608 by a Dutch optician called Hans Lippershey. In 1610, Italian astronomer and physicist Galileo Galilei built an improved telescope based on Lippershey's invention and turned it toward the heavens. In 1668, Isaac Newton invented the reflecting telescope.

further ☆ Gather a couple of lenses from old cameras or magnifying glasses,
studies and see if you can make a refracting telescope. If you prefer, you
can purchase an assortment of lenses from scientific supply houses
such as Edmund Scientific. You can make two cardboard tubes,
one to fit inside the other, from stiff cardboard or mailing tubes.
Can you take pictures of the Moon through your telescope?

did you ☆ ❑ That Galileo was able to see mountains and craters on the
know? Moon, the rings of Saturn, and four of the 12 moons of Jupiter
with his crude telescope.
❑ That a reflecting telescope completed in 1948 at Palomar
Observatory in California used a reflecting mirror 200 inches
(5.08 meters) in diameter.
❑ That in April 1990, after 20 years of planning, the Hubble
Space Telescope was launched into space to see 10 times deeper
into the universe than ever before. At first problems occurred,
but repairs were made by astronauts in December 1993.

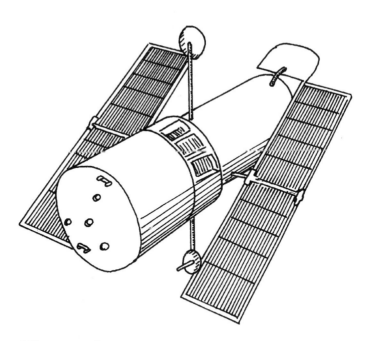

The Hubble Space Telescope

⑨
When was the astrolabe first used?

materials ☆
- ❐ Pencil
- ❐ White poster board, about 5×5 inches (12.7×12.7 cm)
- ❐ Protractor
- ❐ Scissors
- ❐ String, about 8 inches (20.3 cm) long
- ❐ Paper clip
- ❐ Plastic drinking straw
- ❐ Small weight (sinker, washer, etc.)
- ❐ Transparent tape
- ❐ Clear night sky

1. Draw a line ½ inch (1.27 cm) from the top and left side of the poster board. Place the center of the protractor on the point where the lines come together, and mark off the angles from 0 to 90 degrees. Make small lines and number every 10 degrees: 10, 20, 30, etc. Make a smaller line halfway between these numbers to represent every 5 degrees: 5, 15, 25, etc.

procedure ☆

2. Make a small hole where the center of the protractor was located in the preceding step—where the two ½ inch (1.27 cm) lines meet—and thread one end of the string through this hole. Tie the end to the paper clip to keep it from slipping back through. Adjust the length of the string so that the weight swings in an arc about ½ inch (1.27 cm) below the numbers of the degrees. Tie the weight to the free end of the string.
3. Use the string as a guide to trim off the excess part of the card.
4. Tape the straw to the top edge of the card. You will use the straw to sight through. If you aim at the horizon, the string should be at 0 degrees. If you aim at something directly overhead, the string should be at 90 degrees. On a clear night, sight through the straw at the North Star and read the degrees.

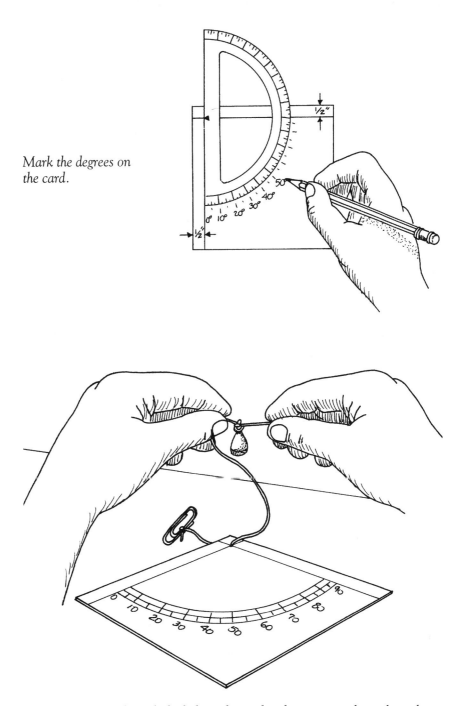

Mark the degrees on the card.

Thread the string through the hole in the card and tie a paper clip and weight to either end.

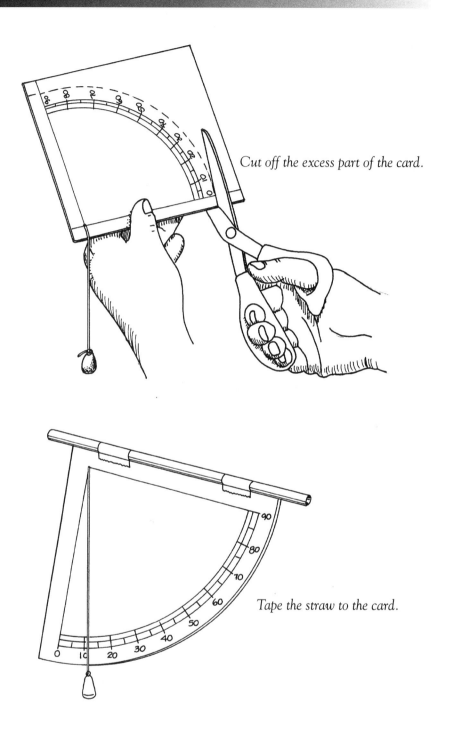

Cut off the excess part of the card.

Tape the straw to the card.

results ☆ The string marks the altitude of the North Star, which is also the latitude of your location.

further ☆ After finding the altitude of the North Star, look at a map to see if
studies the degrees are the same as the latitude where you live. What should your astrolabe read if you sight the North Star from the equator or from the North Pole?

did you ☆ ❒ That the astrolabe is an instrument that ancient astronomers
know? used to measure the angles of celestial bodies above the horizon.
❒ That astrolabes were used in the Middle Ages to determine the positions of heavenly bodies, height of buildings, latitudes, and time.

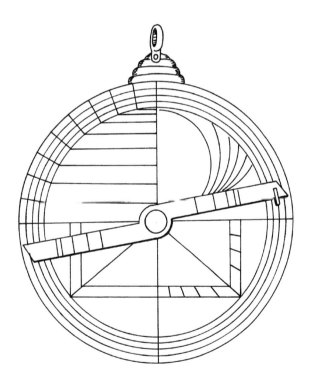

An early astrolabe

10
When can
you see Venus?

materials ☆ ❐ Clear morning or evening sky

procedure ☆ 1. Face the east just before sunrise or the west just after sunset.
2. Find what appears to be the brightest star.

results ☆ The brightest star you see should be the planet Venus. Venus always appears near the Sun. Sometimes you cannot see Venus because it is too close to the Sun.

Caution: Never look directly at the Sun. The Sun is the brightest object in our sky. The Moon and then Venus are the next two brightest objects. While Venus is often called the morning or evening star, it is really not a star because it gives off no light of its own. The light we see from Venus, or any planet, is the light of the Sun reflected off the planet's surface.

further studies ☆ Look up planets in an encyclopedia and learn how many there are. Is Earth a planet? How many planets can you see with the naked eye? Which are the closest and the farthest planets from the Sun?

did you know? ☆ ❐ That about 75 percent of the Sun's rays are reflected from Venus because it is covered with clouds. Because Earth has less cloud cover than Venus, Earth reflects only about 40 percent of the Sun's light.
❐ That Jupiter is more than 1300 times as large as Earth, but Mercury is only about one-sixteenth the size of Earth.
❐ That a 100-pound person on Earth would weigh about 264 pounds on Jupiter.

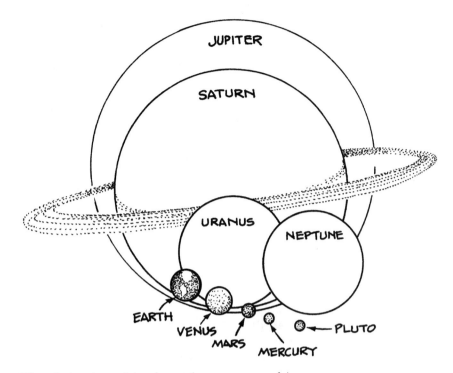

The relative sizes of the planets (not to exact scale)

11
When can
you see Orion?

materials ☆ ❒ Clear winter night (January or February, between 8 and 10 o'clock)
❒ Sky map showing the constellation Orion

procedure ☆ 1. Face south and look at a point about 45 degrees above the horizon. This point is about halfway between the horizon and straight overhead.
2. Find the constellation Orion. Orion can be identified by four major stars with a line of three stars running at an angle near the center. Locate the two brightest of the four major stars. Do they have different colors? Does your sky map give them names?

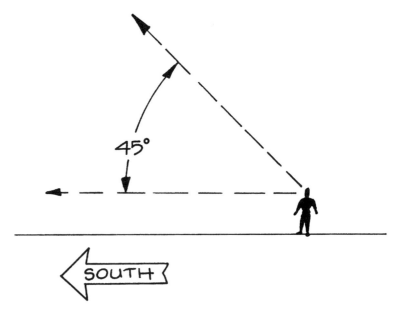

Look in the southern sky to see Orion.

The constellation Orion

results ☆ You have located the constellation Orion (the Great Hunter). You found the two brightest stars, which are called Betelgeuse and Rigel. Betelgeuse is in the northeast corner of the four stars and marks the right shoulder of Orion. Rigel is at the southwest corner marking the left knee. You might have noticed that Betelgeuse has a red tint and Rigel is blue-white.

further studies ☆ What do the three stars in a line represent on Orion? If you watched the constellation through January and February, in what direction do the stars move? Can you find other constellations, such as Taurus (the Bull) or Gemini (the Twins) near Orion?

did you know? ☆ ❑ That in 360 B.C. some people believed the Earth was motionless and at the center of the universe. By 150 B.C. some people thought that the Earth moves around the Sun. In 1554, the Polish astronomer Copernicus published a book that supports the idea that Earth and other planets revolve around the Sun.

❏ That Betelgeuse expands and shrinks from a diameter of about 287 million miles (462 million kilometers) to about 397 million miles (639 million kilometers). These changes cause the brightness of Betelgeuse to vary as well. Betelgeuse has a surface temperature of about 5500 degrees F (3000 C), and it is about 300 light-years away from Earth.

❏ That Rigel is at least 15,500 times brighter than the Sun and that Rigel has a diameter 35 times greater than that of the Sun. Rigel is about 540 light-years away from Earth.

The constellation Taurus

The constellation Gemini

Part 3
The young chemist

Chemistry is the science that deals with the composition, or makeup, of something and the reactions that produce changes in these materials. Chemical change is all around us in the form of iron that rusts, coals burning in furnaces and turning to ash, water evaporating, and gases arising from decomposing materials. It is important to understand and learn how to control these changes so that new materials and new forms of energy might be produced. New alloys, plastics, and other materials are being developed for everyday use, as well as for new materials to be used in space. Many forms of fuels and types of energy would not be available without chemistry.

Our lives would be very different today without chemists and chemistry. Chemistry helps fight disease through the use of drugs and medicines. Even the food we eat is chemically changed inside our bodies to give us energy and to make flesh and bones. Our health has been improved through the discovery of vitamins and how our bodies use and store them.

During the Middle Ages, chemistry was the tool of alchemists, who were mainly concerned with attempting to turn cheaper metals into more valuable gold. Today, chemistry is used by trained people who conduct exciting research and make amazing discoveries.

12
When can you lower the freezing point of water?

materials ☆
- ❏ Two thermometers (Fahrenheit and/or Celsius scale)
- ❏ Two glasses or plastic cups
- ❏ Ice
- ❏ Water
- ❏ Tablespoon
- ❏ Salt
- ❏ Plastic spoon for stirring

procedure ☆

1. Place a thermometer in each glass and fill both glasses with ice. Add a little water to each glass.

Add water to both glasses.

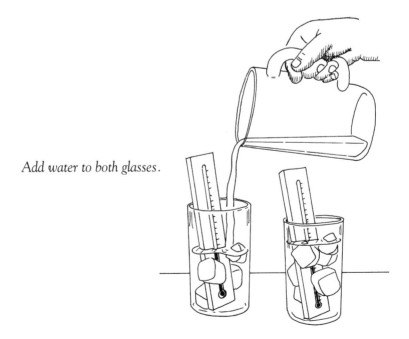

2. Watch the thermometers until both temperatures stabilize. The thermometers should read slightly above 32 degrees Fahrenheit (0 Celsius).
3. Add one tablespoon of salt to one of the glasses and stir the solution with the plastic spoon. Monitor the temperatures.

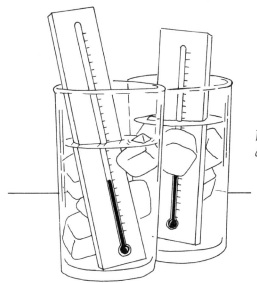

Both thermometers should read about 32 degrees Fahrenheit.

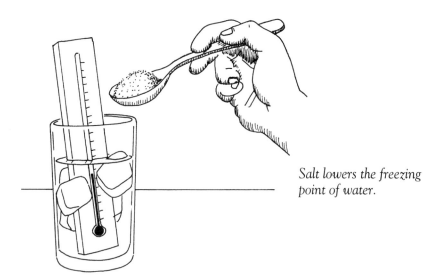

Salt lowers the freezing point of water.

results ☆ The temperature in the glass with salt should drop some. When salt is added, the freezing point of water is lowered to below 32 degrees Fahrenheit (0 Celsius). Salt is added to the ice and water of home ice cream makers to lower the temperature of the water enough to freeze the ice cream mix.

further studies ☆ What other times is salt used to lower the freezing point of water? Why are some northern city streets sprinkled with salt for safer winter driving? Why would salt be one way to remove ice from slippery sidewalks?

did you know? ☆
❏ That sodium chloride is common table salt.
❏ That salt is a compound produced when an acid and a base neutralize each other.

13
When do stalagmites and stalactites form?

materials ☆
- ❏ Two jars
- ❏ Hot tap water
- ❏ Tablespoon
- ❏ Epsom salts (from grocery store)
- ❏ Sheet of cardboard
- ❏ Heavy cotton string

procedure ☆ 1. Fill each jar about two-thirds full of hot water. Then stir in several tablespoons of Epsom salts into each jar until no more salts will dissolve.

Stir in Epsom salts in each jar.

2. Place the jars a few inches (125 mm) apart on the cardboard. Lower one end of the string into one of the jars and the other end of the string into the other jar.
3. Let the part of the string between the jars sag to form a shallow V about half the distance to the cardboard. Set the experiment aside for a few days.

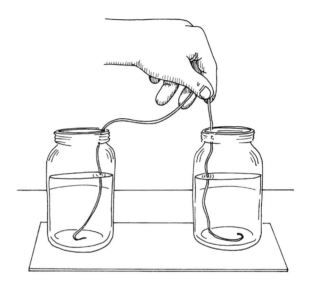

Lower the ends of the string into the jars.

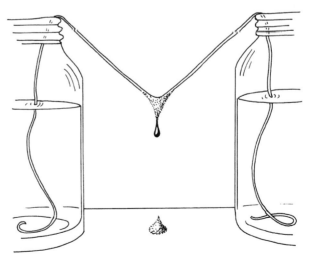

Let the string sag into a V.

results ☆ After a few days, you should find a thin formation growing down from the string and another growing up from the cardboard. In a cave, drops of water containing dissolved limestone fall to the floor. Limestone is made of *calcium carbonate*, which dissolves in the water that flows over limestone rocks. Specks, or crystals, of calcium carbonate, or *calcite*, build up when the water evaporates. This buildup continues year after year to make the formations.

A formation growing down from the roof of a cave is called a *stalactite*. A formation growing up from the floor is called a *stalagmite*. In your experiment, the salt in the solution travels up the string and is deposited where the water drips. The salt remains as crystals after the water evaporates, creating the stalactites and stalagmites.

further ☆ Place a pan of tap water in sunlight and allow the water to
studies evaporate. Can you see chemical deposits in the pan? Wet the inside of a clear glass and allow the water to evaporate. Can you see spots on the glass?

Could this water, slowly dripping over many years, make stalactites and stalagmites?

did you ☆ ❏ That sometimes stalactites and stalagmites join to form
know? columns or stone curtains.
❏ That calcite builds up into colorful formations that look like icicles.

14
When can bones become flexible?

materials ☆
- ❏ Chicken bone (drumstick) or other similar bone
- ❏ Glass or jar
- ❏ Several bottles of vinegar

procedure ☆
1. Place the bone in the glass and fill it with vinegar. The vinegar should completely cover the bone.
2. Keep the bone submerged in vinegar for several days. Replace the old vinegar with fresh every couple of days.

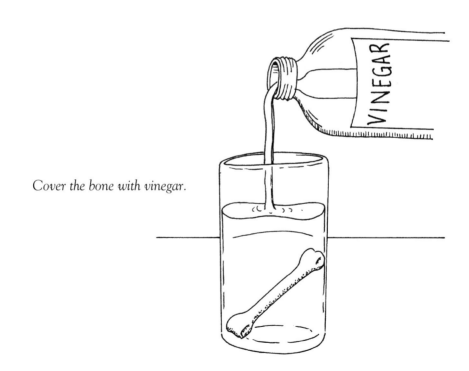

Cover the bone with vinegar.

results ☆ After a few days, the bone should be flexible. Try bending the bone. Bones are hard because they contain *calcium phosphate*. The acetic acid in the vinegar changes the calcium phosphate in the bone to *calcium acetate*. When the bone loses its calcium phosphate, it becomes more flexible.

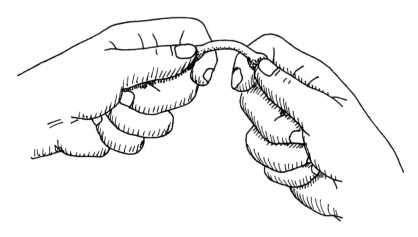

The bone becomes flexible.

further studies ☆ Try larger bones. Can you bend one in a circle or tie it in a knot? To find out what bones are made of, look up "bone" in an encyclopedia.

did you know? ☆ ❏ That about two-thirds of animal bone is mineral matter.
❏ That the mineral matter of animal bones is used for fertilizer.

15
When does
paper contain starch?

materials ☆
- ❑ Tablespoon
- ❑ Water
- ❑ Small glass
- ❑ Iodine
- ❑ Medicine dropper
- ❑ Different types of paper (newspaper, writing paper, paper towels, etc.)

procedure ☆
1. Put two or three tablespoons of water in the glass and add an equal amount of iodine. Stir the solution.
2. Use the medicine dropper to place a few drops of the iodine solution on the paper. If you don't have a medicine dropper, just dip a strip of the paper into the solution.

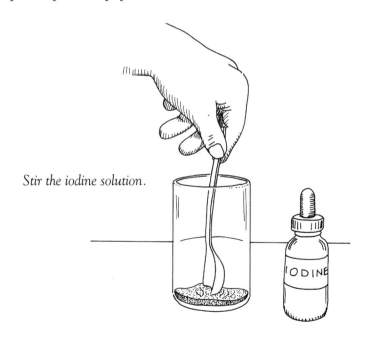

Stir the iodine solution.

Place a drop or two on the papers.

results ☆ If the paper is the same color as the iodine solution or slightly lighter, the paper does not contain starch. If the solution turned bluish-black or black, the paper contains starch. When some papers are made, a film of a starch solution is put on the paper to give the paper a smooth surface and to hold the fibers together. Laundry starch works the same way on clothes.

further ☆ Use the same procedure to test a small piece of white bread for
studies starch. Check slices of a potato and an apple. Drop the solution on a small mound of flour or a small mound of salt.

did you ☆ ❑ That starch is one kind of food made by plants that contain
know? *chlorophyll*, the green coloring matter.
❑ That as early as 184 B.C., starch was prepared from cereal grains. It was manufactured, first from potatoes and then from corn, in the United States in the early 1800s.

Part 4
The young meteorologist

Meteorology is the study of the weather and the atmosphere. It attempts to explain why weather conditions occur. The study of meteorology requires using other sciences, such as physics, chemistry, and mathematics. Meteorologists use physics to explain the movement of the atmosphere. In addition, physics helps us understand what causes rain, snow, and hail to form, and what causes lightning and other electrical phenomena.

Meteorologists use chemistry to study the gases that make up the air we breathe and to study the impurities that pollute it. Mathematics is used to accurately calculate the speed of storms, to understand what makes the wind blow, and to make more accurate weather forecasts.

Meteorologists try to learn as much about the atmosphere as they can. They use *thermometers* to measure the temperature of the atmosphere, *barometers* to measure atmospheric pressure, and *hydrometers* to measure its moisture content. *Rain gauges* are used to determine the amount of rainfall, and *anemometers* are used to measure the speed and direction of the wind.

Pilots, ship captains, farmers, and highway maintenance departments are just a few of the people that are concerned with weather conditions. Large gas and electric companies use weather forecasts to determine estimates of their needs in advance. Even something as advanced as the space shuttle has had launches delayed by weather.

Meteorologists also can help us decide what clothes we should wear or when to plan the first ski trips of the season. It is easy to see that how we live, where we live, and what we do is, in some way, dependent on weather conditions and the science of meteorology.

16
When was the thermometer invented?

materials ☆
- ❑ Clear glass bottle, pint or quart (0.47 or 0.95 liter)
- ❑ Water
- ❑ Medicine dropper
- ❑ Red food coloring
- ❑ Plastic drinking straw
- ❑ Rubber stopper with hole or Plasticene clay (used around air-conditioners)
- ❑ Few drops of cooking oil
- ❑ A 3-x-5-inch (76.2-x-127-mm) card
- ❑ Adhesive tape
- ❑ Thermometer for calibration
- ❑ Pencil

procedure ☆
1. Fill the bottle with water and use a medicine dropper to drop in a few drops of food coloring. Use the straw to stir until the color is uniform throughout the bottle.
2. Push the straw through the hole in the rubber stopper. Press the stopper into the opening in the bottle (or mold the clay around the straw in the opening of the bottle). The straw should stick down into the water a couple of inches.
3. Use the medicine dropper to add more colored water to the opening in the top of the straw. Fill the straw to about one-fourth of the way up. Add a couple drops of cooking oil to keep the water from evaporating.
4. Fasten the card to the back side of the straw with tape.
5. Use another thermometer to determine the temperature in the room. Mark this temperature at the level of the water in the straw. Calibrate your thermometer by placing both thermometers in cooler and warmer locations and marking the degrees on the card.

Add food coloring to the water.

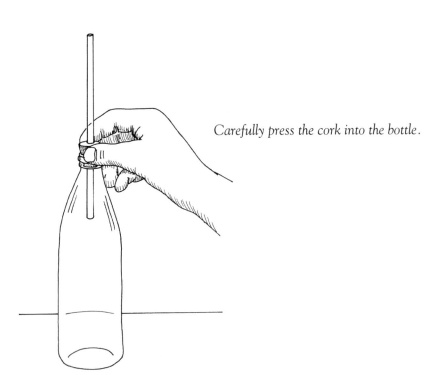

Carefully press the cork into the bottle.

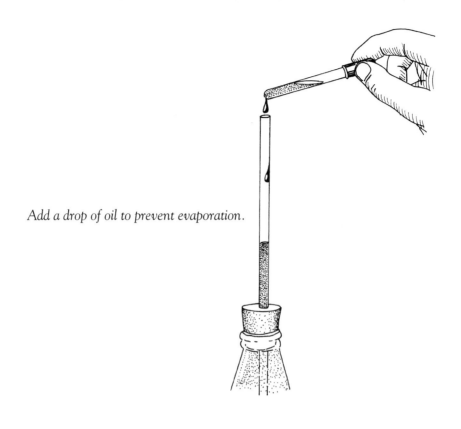

Add a drop of oil to prevent evaporation.

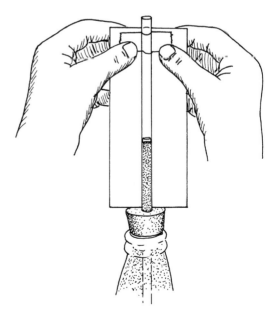

Tape the card to the straw.

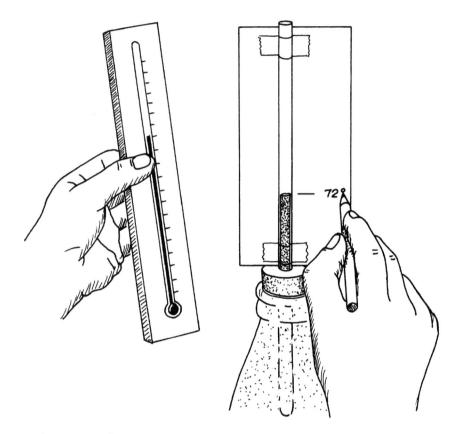

Calibrate your thermometer.

results ☆ The water in your thermometer should rise and fall slightly as the temperature changes. The water level changes because most liquids expand when heated and contract when cooled.

further studies ☆ Mark one side of your thermometer in Fahrenheit and the other in Celsius. At what Celsius temperature does water freeze? At what Celsius temperature does water boil? How do these temperatures compare to Fahrenheit? Is Celsius used for scientific measurements?

did you know? ☆
❏ That the first thermometer was made in 1593 by the astronomer Galileo.
❏ That in 1714 Gabriel Fahrenheit made a mercury thermometer of the type we use today.
❏ That Anders Celsius introduced the Celsius scale in 1742.

17
When was the barometer invented?

materials ☆
- ❏ Four paper clips
- ❏ Tall clear glass or jar
- ❏ Water
- ❏ Red food coloring
- ❏ Bowl
- ❏ Marking pen

procedure ☆ 1. Slide the paper clips onto the rim of the glass. Space them equal distances around the rim and press them down on the glass.

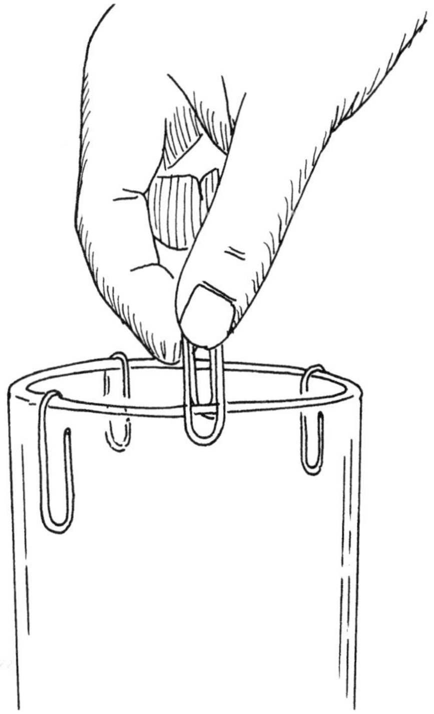

Place paper clips around the rim.

2. Fill the glass about two-thirds full of water and stir in a few drops of food coloring.
3. Place the bowl upside down over the glass. Then turn the bowl and the glass of water rightside up. Some of the water runs into the bowl, and the rest of the water remains in the glass. Add a little more colored water to the bowl to fill it about halfway.
4. Mark the level of the water in the bowl and place it in a room where the temperature will not change much. Temperature changes can cause the air in the glass to expand or contract and give false readings.

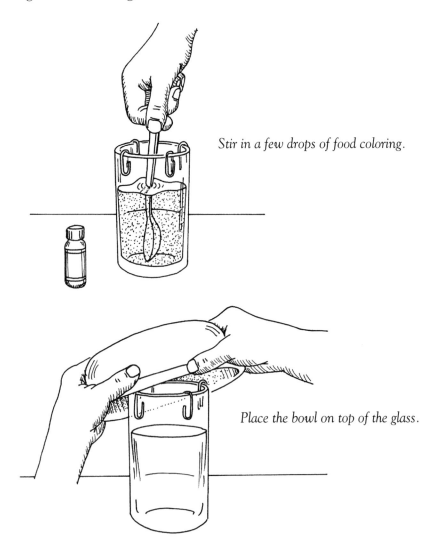

Stir in a few drops of food coloring.

Place the bowl on top of the glass.

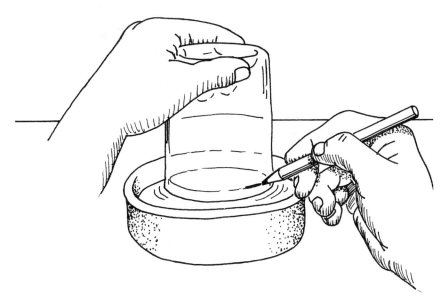

Mark the water level.

results ☆ As the weather changes, the atmospheric pressure changes. This causes the water in the glass to rise and fall. If the water level drops, it indicates a low-pressure area; a storm is approaching. If the water level rises, it could indicate a high-pressure area and probably means fair weather. Meteorologists refer to areas of different air pressures as *highs* or *lows*. A barometer is simply an instrument that measures these air pressures. A barometer that is commercially made is much more accurate than your homemade one, but your barometer demonstrates the principle of how they work.

In 1646 Otto von Guericke, a German scientist, invented a water barometer made of a brass tube 34 feet (10.4 meters) long with a closed section at the top. He attached the barometer to the side of his house and floated the figure of a little man inside the tube. People could see the figure floating high in good weather and floating lower when it was stormy.

further ☆ Today's barometers are much more accurate than Otto von
studies Guericke's water barometer, but they still operate on atmospheric pressure. Can a barometer be used to measure the height of a mountain? Airplanes have *altimeters* that can be set at ground level before takeoff or to the barometric pressure of their location. Is the

German scientist Otto von Guericke

Italian physicist Evangelista Torricelli

before takeoff or to the barometric pressure of their location. Is the altimeter a type of barometer?

Compare your barometer with the weather map from your local newspaper. Can you forecast a coming storm?

did you ☆
know?

- ❏ That some human body aches and pains occur more often in low-pressure areas than in high-pressure areas?
- ❏ That the lines on a weather map are called *isobars*, from the Greek words *iso* (meaning equal) and *bar* (meaning weight).
- ❏ That in 1643 Evangelista Torricelli, a pupil of Galileo, discovered that the weight of all the air above any point on the earth's surface is equal to the weight of a column of mercury about 30 inches (76.2 cm) high. Today, under normal conditions, the atmospheric pressure is considered to be 29.53 inches of mercury at sea level.
- ❏ That in 1654, to show how strong air pressure can be, Otto von Guericke fastened two halves of a hollow sphere together and made them airtight. He then pumped out the air. The outside air pressure was so great that it took 16 horses to pull the halves apart.

18
When does air rise?

materials ☆
- ❏ Wood pencil (with attached eraser)
- ❏ Paper (8½ × 11 inches)
- ❏ Scissors
- ❏ Metal thimble
- ❏ Needle
- ❏ Electric table lamp with lamp shade
- ❏ Spool (sewing thread)

procedure ☆

1. Ask an adult to assist you with the following procedures. Mark the pattern of a spiral on the paper. Make the turns about 1 inch (2.54 cm) wide. Remembering to always point the scissors away from you, cut the pattern from the paper, but leave enough space in the center to partially insert the thimble.

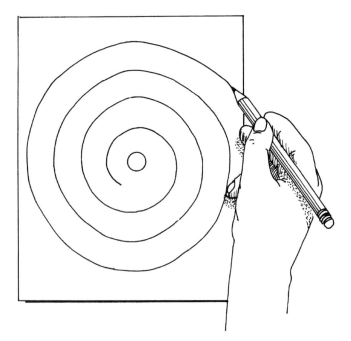

Mark a spiral on the paper.

2. Make a hole in the center of the spiral and press the thimble part way through.
3. Carefully insert the needle upside down, and straight, into the pencil's eraser.

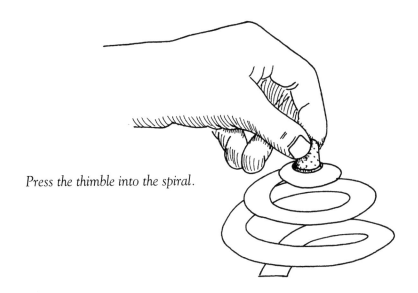

Press the thimble into the spiral.

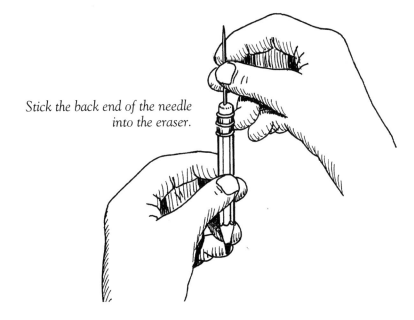

Stick the back end of the needle into the eraser.

4. Remove the threaded nut from the top of the lamp shade and place the spool over the threaded stud. Now place the pointed end of the pencil into the hole in the spool.
5. Place the open end of the thimble, in the spiral, on the point of the needle. Turn on the lamp.

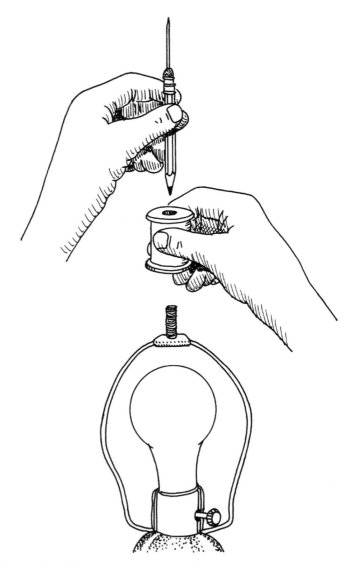

Place the spool and the pencil on top of the lamp.

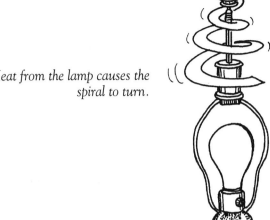

Heat from the lamp causes the spiral to turn.

results ☆ After a few minutes, the spiral should begin to turn. The lamp's lightbulb heats the air and the air expands, making the air near the lightbulb lighter than the surrounding air. Cooler, heavier air moves in and pushes the warm air up. The warm air pushes on the spiral and makes it turn.

further studies ☆ Blow up a balloon, measure the diameter, and then place the balloon in the refrigerator. After several minutes, remove the balloon and measure it again. Has the air in the balloon shrunk? At room temperature, do you think the balloon will get larger? The balloon shrinks in size when placed in a cooler environment because the air molecules inside the balloon are moving closer together.

did you know? ☆
❐ That air over the equator is always rising.
❐ That because of gravity, air pressure is greatest at ground level and gets less the higher you go.
❐ That rising columns of air are called *thermals* and that they allow soaring birds to remain aloft for hours without flapping their wings.

19
When does air sink?

materials ☆
- ❐ Match
- ❐ Small slip of paper
- ❐ Large, thick-glass jar with lid
- ❐ Refrigerator (freezer section)

procedure ☆

1. Ask an adult to assist you with the following procedures. Carefully light the paper and quickly blow it out. You want to capture the smoke. Drop the paper in the jar and replace the lid.

Capture some smoke in a jar.

2. Place the jar in the freezing compartment of the refrigerator. Let it set about 10 minutes.
3. Remove the jar from the freezer and take it to a room without drafts. Remove the lid and observe the smoke. Turn the jar upside down. Observe the smoke.

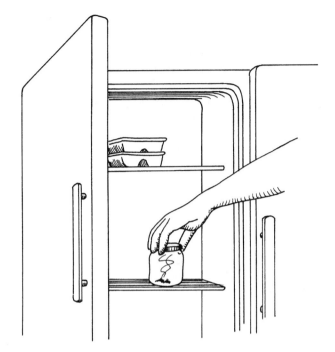

Cool the air in the jar.

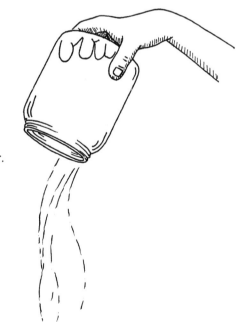

Pour the air from the jar.

results ☆ When you open the jar rightside up, little of the smoky air is able to rise from the jar. But when you turn the jar upside down, the air sinks toward the floor. The cold air in the jar is denser than the warmer air in the room. The denser the air, the heavier it is.

further studies ☆ Open the refrigerator door slightly and feel the air near the bottom. Is cold air sliding out? Is warm air going in the top of the door? Why would you expect the air to be constantly sinking over the North Pole and the South Pole? How does this cause trade winds?

did you know? ☆
❏ That the Antarctic—with temperatures falling well below –60 degrees F (–50 degrees C)—is the coldest and windiest place on Earth.
❏ That the world's first weather satellite was launched on April 1, 1960, from Cape Canaveral. It was named TIROS (an acronym from television and infrared observation satellite).

20
When can you see your breath?

materials ☆ ☐ Drinking glass
☐ Ice cubes

procedure ☆ 1. Fill the glass about half full of ice cubes.
2. Gently blow over the rim of the glass.

Blow across the top of the glass.

results ☆ You should be able to see your breath blowing across the top of the glass. The air you breathe out from your lungs is warm and moist. The ice cubes cool the air in much the same way as if you were outside on a cold day. Cold air cannot hold as much moisture as warm air. Therefore, some of the moisture in your breath condenses into tiny droplets that you can see. Condensation occurs when a gas—water vapor in this case—is cooled enough that it changes back into a liquid.

further studies ☆ How can warm, moist air rising into a cooler atmosphere create clouds? If enough moisture condenses, will it rain? If the cloud is high enough, could it hail? Can you think of three forms of water in the air?

did you know? ☆
- ❏ That if all the water in the air fell at the same time, it would cover the entire earth with about 1 inch (2.54 cm) of water.
- ❏ That no new water is ever made. In the water cycle, the earth's water is used over and over again. The rain we see has fallen millions of times before.
- ❏ That about 70 percent of the earth's surface is covered with water.

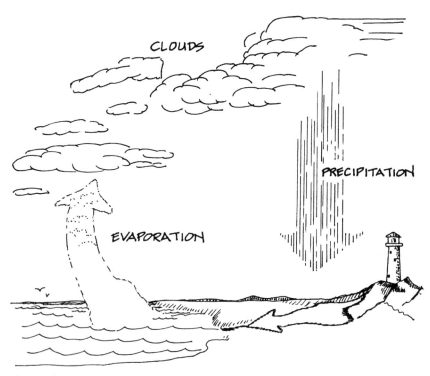

The earth's water cycle

21
When does water get into the air?

materials ☆
- ❏ Two jars the same size (one with lid)
- ❏ Water
- ❏ Marking pen

procedure ☆
1. Fill both jars about half full of water. Screw the lid on one of the jars and leave the other one open.
2. Mark the water levels of each jar.
3. Let the jars stand side by side for several days. Monitor the water levels and mark any changes that occur.

Mark the water levels.

Water evaporates from the open jar.

results ☆ The water level in the open jar keeps falling. Water does not evaporate at the same rate every day; it depends on the temperature and the amount of water as humidity in the air. The warmer the air, the more moisture it can hold. The air in the sealed jar becomes saturated, and when the air is saturated, no more water can evaporate in that air.

further ☆ Which jar has the air that contains the most water vapor? Is this **studies** air saturated? When we exercise, do our bodies give off moisture? Why are we more uncomfortable on warm, humid days?

did you ☆ ❑ That a cubic foot of saturated air at 90 degrees F (32 degrees C) **know?** contains five times as much moisture as saturated air at 40 degrees F (4 degrees C).
❑ That when the weather station says the relative humidity is 50 percent, this means that the air contains only half of the moisture it can hold.

22
When does solar energy change the seasons?

materials ☆
- ☐ Flashlight
- ☐ Piece of paper
- ☐ Ruler

procedure ☆
1. Turn the flashlight on and aim it straight down at the paper. Measure the diameter of the circle of light on the paper.
2. Hold the flashlight at an angle and aim it at the same spot. Measure the diameter of the lighted area.

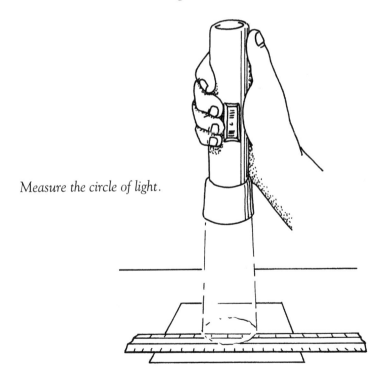

Measure the circle of light.

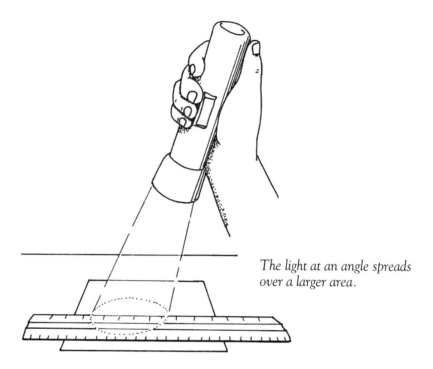

The light at an angle spreads over a larger area.

results ☆ When the light comes from directly overhead, the energy is concentrated in an area. But when striking the paper at an angle, the same amount of energy is spread over a larger area. The angle that the Sun's rays strike Earth is a large part of what determines the temperature. Sunshine strikes Earth near the equator from directly overhead (or at about 90 degrees). Near the polar regions, the rays come in more slanted. This slanting is what causes the difference in temperatures between the poles.

If Earth's axis were straight up and down, instead of being tilted 23.5 degrees, the seasons would not change. With a tilt of 23.5 degrees, about half of the time the North Pole is tilted toward the Sun while the South Pole is tilted away. This tilting of the axis causes the four different seasons.

The result is spring and summer in the Northern Hemisphere and at the same time autumn and winter in the Southern Hemisphere. During the other half of the year, the North Pole is tilted away from the Sun and the seasons reverse. As Earth orbits around the Sun, its axis is always tilted 23.5 degrees.

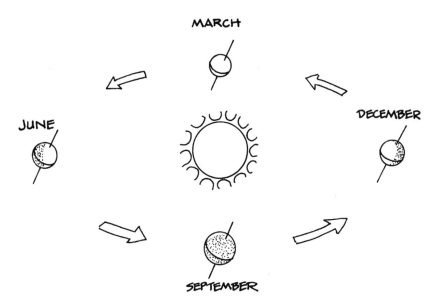

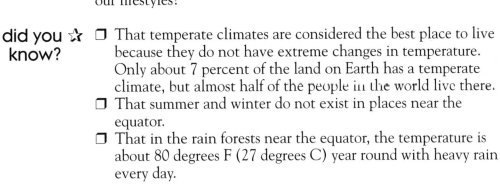

The tilt of Earth causes a change in temperature on its surface.

further studies ☆ Look at a map of the world and find the most populated areas. What type of weather do they have? How do seasons affect the kind of work we do? How much do we depend on agriculture for our lifestyles?

did you know? ☆
- ❏ That temperate climates are considered the best place to live because they do not have extreme changes in temperature. Only about 7 percent of the land on Earth has a temperate climate, but almost half of the people in the world live there.
- ❏ That summer and winter do not exist in places near the equator.
- ❏ That in the rain forests near the equator, the temperature is about 80 degrees F (27 degrees C) year round with heavy rain every day.

23
When does fog form?

materials ☆   ❏ Clear glass jar
           ❏ Hot tap water
           ❏ Tea strainer
           ❏ Ice cubes

procedure ☆   1. Fill the jar about half full of hot water and place the tea strainer over the opening.
            2. Fill the tea strainer with ice cubes and observe the air inside the jar.

Place the strainer on the jar.

Place ice cubes in the strainer.

results ☆ Warm air holds more water than the same volume of cold air. The moist air condenses into a mist when the air is cooled below the dew point. The *dew point* is the temperature at which air has all of the water it can hold and the water vapor condenses. If it forms a mist, it is called *fog*, if it condenses on grass or sidewalks, it is called *dew*.

further studies ☆ Fog forms along the coasts when warm air from the land drifts over cold seas. Does this fog affect shipping? Can fog close an airport? Fog reduces visibility. Why can this cause accidents on highways?

did you know? ☆ ❑ That in larger cities air is being polluted by exhaust fumes from cars, trucks, and buses. When this smoke mixes with fog and reacts with sunlight it is called *smog*. Smog can damage people's health, ruin crops, and even erode stone buildings.
❑ That the term "fog as thick as pea soup" came about when London, England, had terrible, thick smog colored green by the smoke from factories and coal fires.

24
When do off-shore breezes blow?

materials ☆ ❏ Two thermometers
❏ Large jar of water at room temperature
❏ Large jar of dirt

procedure ☆ 1. Place one of the thermometers in the jar of water and insert the other thermometer into the dirt in the other jar.
2. Place the two jars side by side in the sunlight and monitor the temperatures for several minutes. Check the temperatures after the Sun goes down. How do the temperatures compare?

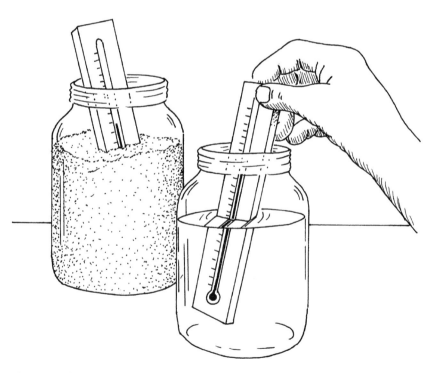

Place one thermometer in water.

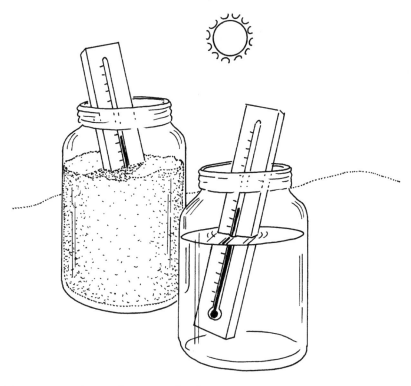

Temperature changes faster over land.

results ☆ In sunlight, the temperature in the dirt starts to rise faster than the temperature of the water. At night, the temperature of the dirt drops faster than the water. Soil absorbs heat faster and loses it quicker than water. Near the beach during the day, the ground warms the air and causes the air to rise. Cooler air from the sea comes in. At night, the land becomes cooler than the seawater and the air over the water rises while the air over the land flows out.

further studies ☆ Because warm air rises and cooler air rushes in to replace it, does this mean that air flows from high-pressure areas to low-pressure areas? With winds normally moving from west to east, explain how highs and lows move on a weather map.

❏ That the *trade winds* are steady winds flowing toward the equator. Early sailors used these constant winds as navigational aids. Columbus might never have reached America without the trade winds.

❏ That for centuries, sailors lost at sea looked for clouds to guide them to land, because low clouds often form over islands.

Part 5
The young biologist

Biology is the study of the origins, physical characteristics, habits, and life processes of living things. One of the differences between living things and nonliving things is that living things take in and use food. While living things grow and produce more of their own kind, they also are made up of the same chemical elements as nonliving things.

When the compositions of living things are analyzed by chemists, the chemists identify elements commonly found in the earth, in sea water, and in the air. These *elements* usually come together to form *compounds*. Water, for example, is a compound of hydrogen and oxygen and makes up between 65 and 90 percent of the weight of most living things.

As equipment and instruments become more advanced, scientists can probe deeper into this curious phenomenon called life. Life is one of the great mysteries of the universe.

25
When can you detect your pulse?

materials ☆
- ❏ Wooden match
- ❏ Thumbtack
- ❏ Table

procedure ☆

1. Ask an adult for permission to use a match. Insert the point of the thumbtack into the end of the match to form a base.
2. Place your finger on your wrist, and find the place where you can feel your pulse.
3. Lay your arm, with the wrist up, on a table or other solid object. Stand the match on this point and observe any motion at the top of the match.

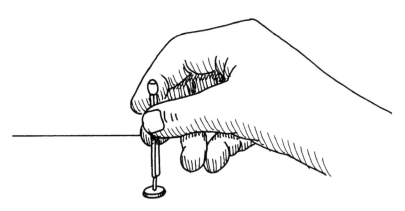

Press the match onto the tack.

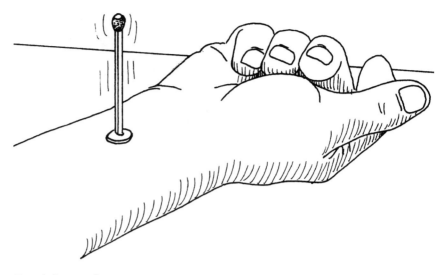

Stand the match on an artery.

results ☆ You should see a slight, but steady, movement at the top of the match. This movement reflects the pulsing of your blood through the artery in your wrist.

further ☆ Let your arm hang down limp for several seconds, and then look
studies closely at the top of your hand. You should see a few blue lines just under the skin. These are blood veins (not arteries). Push your finger along one of the veins toward the knuckles. Notice how the vein seems to disappear. Release your finger and the vein instantly fills back up. The *veins* have valves in them that allow the blood to flow steadily in one direction only: to the heart. *Arteries* carry blood in pulses away from the heart. Can you feel a *pulse* in a vein? What are *capillaries* and what do they do? Is your skin full of capillaries?

did you ☆ ❏ That the circulatory system extends throughout the body.
know? Blood carries food and oxygen throughout the body and, on the return trip, removes nitrogen wastes and carbon dioxide.
❏ That about 250 million of the cells in the blood are destroyed and replaced every day.
❏ That the heart keeps the blood moving so fast that all the blood in the body passes through the heart every two or three minutes.

26
When was the microscope first used?

materials ☆ ❒ A strong convex lens (from an old camera or from a scientific supply house such as Edmund Scientific)
❒ Magnifying glass

procedure ☆ 1. Place the object you want to look at about 2 feet (60 cm) from your eyes.
2. Place the convex lens slightly above the object.
3. Hold the magnifying glass close to one eye and adjust the focus.

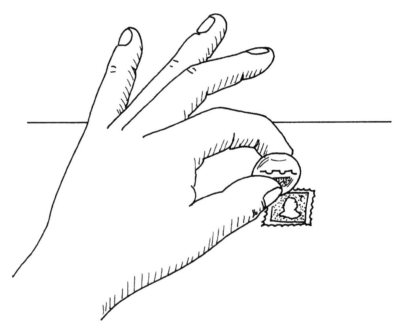

Place the lens just above the object.

results ☆ You should see an enlarged image of the object. A magnifying glass is a single convex lens and is a simple form of a microscope. But better microscopes are made up of several lenses for increased magnification. In your microscope, the lens just above the object is called the *objective lens*. The lens close to your eye is called the *eyepiece*.

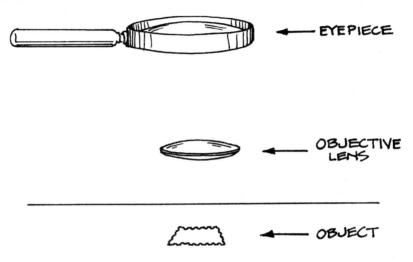

A simple form of microscope

The type of lenses we use today were introduced in the late 1500s. Usually, credit is given to Dutch spectacle-maker Zacharis Janssen for discovering the principle of the microscope in about 1590. Robert Hooke, an English scientist, constructed a multiple-lens microscope illuminated with an oil lamp. He discovered that plants and animals were made up of tiny cells, and he published his results in 1665. C.A. Spencer made the first American microscopes in Canastota, New York, in about 1838.

did you ☆
know?
- ❒ That engravers probably used glass globes filled with water as magnifying glasses 3000 years ago.
- ❒ That electron microscopes use streams of electrons to magnify objects too small to be seen by a regular microscope.

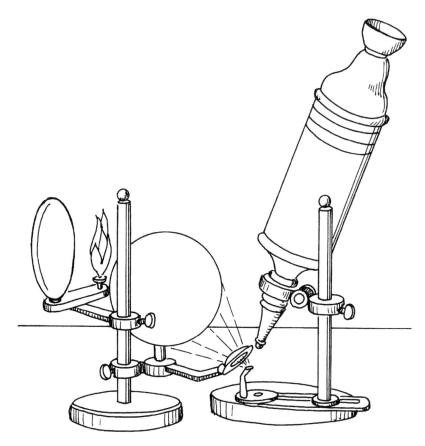

Robert Hooke's compound microscope

27
When don't your taste buds work?

materials ☆ ❑ A small amount of dry, instant coffee

procedure ☆ 1. Hold your nose and place a few grains of instant coffee on your tongue. Continue holding your nose for a few seconds.
2. Spit out the grains, release the grip on your nose, and inhale.

results ☆ When you place the grains on your tongue, they quickly start to dissolve, but you probably cannot taste them. After you release your nose and inhale, you quickly taste the coffee. Nerves from the taste buds carry signals to the taste center in the brain. The taste center is close to the *olfactory* (smell center). The sense of taste and smell are closely related. It is often difficult to tell the difference between the two. This closeness is why when you have a cold, food seems tasteless.

Nerves carry signals to the taste center.

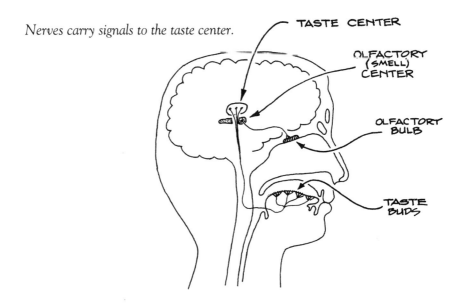

TASTE CENTER

OLFACTORY (SMELL) CENTER

OLFACTORY BULB

TASTE BUDS

further ☆ studies

Taste is one of the five senses. The sense of taste is made up of four reactions: sweet, sour, salty, and bitter. Other tastes are simply combinations of these four. Tastes, like odors, are caused by chemicals dissolved in liquids around the sense organs.

Place a few grains of dry sugar or salt on the tip of your tongue. How long does it take for you to taste it? Does it have to dissolve first? Now place the grains on the center of your tongue. Does it take even longer to taste the grains? Is the tip of your tongue more sensitive to sweet and salty tastes than the center? Compare tastes of different kinds of cold drinks. Can you tell the difference blindfolded?

did you ☆ know?

❏ That different parts of the tongue are sensitive to different tastes.
❏ That taste in animals plays an important part in their selection of the right foods. If their bodies need salt, for example, they develop a taste for it.
❏ That the same drink, depending on whether it is hot or cold, can taste different.

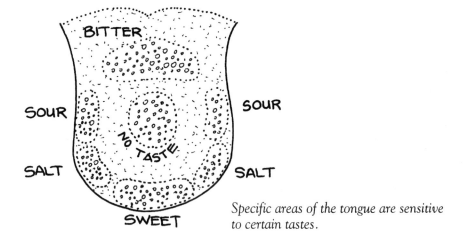

Specific areas of the tongue are sensitive to certain tastes.

28
When do leaves fall?

materials ☆ ❏ Tree leaves (maple, oak, or ash) in autumn after they have
fallen

procedure ☆ **1.** Look closely at one of the leaves. Notice that it has two main
parts: the *blade* and the *stem* or *petiole*. The blade is the broad
part of the leaf that holds the green food-making cells. The
stem holds the leaf to the plant, but it also brings water to the
leaf and carries liquid food back to all parts of the plant. Notice
that the leaf is no longer green.

The two main parts of a leaf.

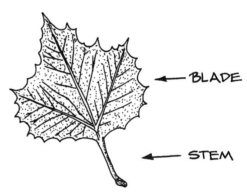

← BLADE

← STEM

The cross section of a stem.

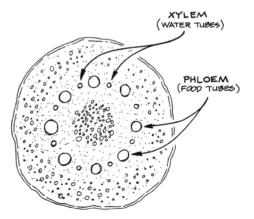

XYLEM
(WATER TUBES)

PHLOEM
(FOOD TUBES)

CROSS SECTION of STEM

results ☆ During the summer, when the leaf is fully grown, it produces large amounts of food. As summer continues, the young leaf begins to turn from a bright green to a darker green. Then something strange begins to happen at the base of the stem. A ring of cells, called *abscission cells*, begins to turn into cork. In late summer and early fall, these corky cells grow across the stem and slowly block the tiny tubes that carry water and food to and from the blade.

By late summer to early fall, the water supply is completely cut off and the leaf stops making food. With its food supply cut off, the leaf loses its green color and reveals its hidden colors of yellow, red, and orange or purple. The leaf continues to hang on until the stem breaks off cleanly through the abscission (cork) cells and flutters to the ground.

further ☆ Study the leaves of trees in your area. Are they broad-leaf trees,
studies needle-leaf trees, or both? All leaves use air, water, and sunlight to make food for the plant. Can a leaf be thought of as a small factory? Do needle-leaves make food just as other leaves do? What gives the leaves their green color? What is *photosynthesis*?

did you ☆ ❐ That one kind of South American palm tree has a leaf over 26
know? feet (7.9 meters) long and 5 feet (1.5 meters) wide. It has the world's biggest leaf.
❐ That a single leaf of the giant water lily, *Victoria regia*, named after Queen Victoria, can grow to be 6 feet (1.8 meters) across, and is strong enough to support a boy or girl?

29
When do plants give off moisture?

materials ☆
- ☐ Pencil
- ☐ A piece of thin cardboard
- ☐ A thick leaf with a strong stem (such as philodendron)
- ☐ Modeling clay
- ☐ Two drinking glasses
- ☐ Water

procedure ☆

1. Use the pencil to make a small hole in the center of the cardboard. Push the leaf stem through the hole until the leaf almost rests on the cardboard. The stem should stick down a few inches (12.7 cm) below the cardboard.

Make a small hole in the card.

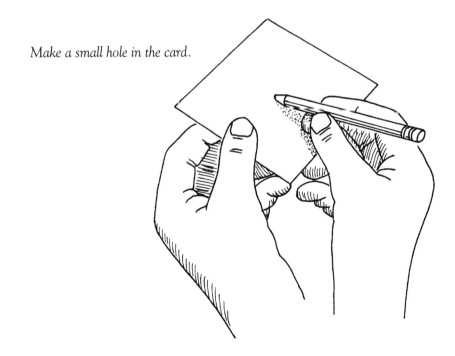

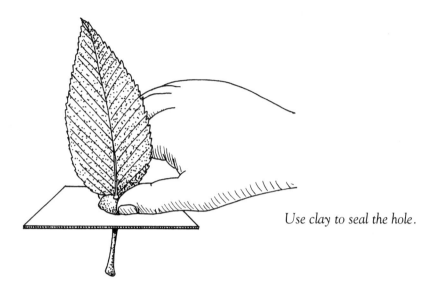

Use clay to seal the hole.

2. Press small pieces of clay against the cardboard and around the stem to seal the hole.
3. Fill one of the glasses with water and place the card on top of the glass. The cardboard—with the leaf on top—should completely cover the glass, and the stem should be in the water.

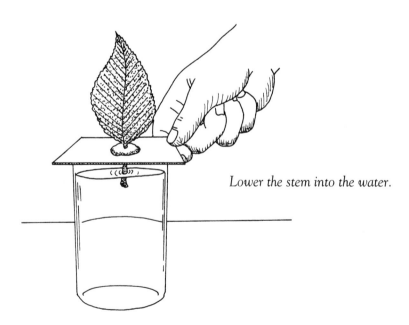

Lower the stem into the water.

4. Put the empty glass upside down over the cardboard and cover the leaf. Place the glasses in normal sunlight for a few hours and observe the empty glass.

Place the glass in sunlight.

results ☆ Small drops of moisture begin to appear on the inside of the top glass. The leaf gives off water that it has drawn up through its stem from the glass of water. This process is called *transpiration*. It is similar to perspiration, the production of sweat, in animals. Plants give off water mostly through the tiny openings (*stomata*) on the lower surface of the leaves. The amount of water they give off depends largely on how much was soaked up by the roots of the plant. Leaves make food only in daylight, but water flows in and out of leaves day and night.

further studies ☆ Study the leaves of a tree or leafy bush. Notice that the stems grow extra long if the branches have to hold leaves out to reach sunlight. But the stem is short if no other leaves block the sunlight. Stand under a tree and look up at the pattern of leaves. At sometime during the day, the full surface area of each leaf is exposed to sunlight. Do stems also twist to hold broad leafs toward the sunlight?

did you ☆ know?

❒ That the leaves of an elm tree might give off over 200 gallons (757 liters) of water a day.

❒ That the leaves of an acre of uncut grass in a meadow might give off 1200 gallons (4542 liters) of water a day.

30
When did dinosaurs live?

materials ☆
- ☐ Yardstick
- ☐ Table
- ☐ Pencil
- ☐ Strips of paper
- ☐ Thumb tacks

procedure ☆ 1. Place the yardstick on a table. Write "First Reptile" on a slip of paper. Use a thumb tack to fasten the paper on the 33.5 inch (85.0 cm) mark on the yardstick.

Mark the time of the first reptile.

2. Write "Triassic" on a strip of paper and use a thumb tack to stick it to the 24.5 inch (62.2 cm) mark on the yardstick. Write "First Dinosaur" on a strip of paper and stick it to the 22.5 inch (57.1 cm) mark.
3. Write "Jurassic" on a strip of paper and stick it to the 20.5 inch (52.0 cm) mark.
4. Write "Cretaceous" on a strip of paper and stick it to the 14.5 inch (36.8 cm) mark.
5. Write "End of Dinosaurs" on a strip of paper and stick it to the 6.5 inch (16.5 cm) mark.
6. Write "First Humans" and stick it to the 1.5 inch (3.8 cm) mark.

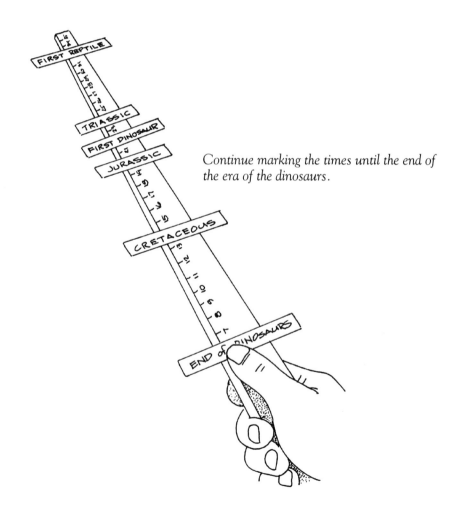

Continue marking the times until the end of the era of the dinosaurs.

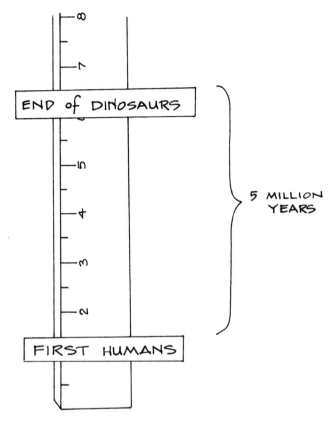

8
7
END of DINOSAURS
5
4
3
2
FIRST HUMANS

} 5 MILLION YEARS

Mark the time when the first humans appeared.

results ☆ Each inch (2.54 cm) on the yardstick represents 10 million years in the history of the earth. The first reptile appeared about 335 million years ago. The Triassic period occurred between 245 to 208 million years ago. All of the land was joined in one continent called Pangaea. The first dinosaurs appeared and flying reptiles took to the air.

Between 208 and 145 million years ago, the Pangaea continent split in two and the Jurassic period occurred. Giant plant-eating dinosaurs (*sauropods*) walked on four legs, and many kinds of meat-eating (*theropods*) dinosaurs developed and walked on their two back legs.

The Cretaceous period lasted between 145 and 65 million years ago. The land divided into several continents. Dinosaurs lived in deserts, warm swamps, and the cold Arctic. Most of the dinosaurs we know come from this time. This includes one of the last meat-eaters and the biggest dinosaur hunter, *Tyrannosaurus rex*. The first humans appeared only 1.5 million years ago.

further ☆ studies

If a yardstick is not handy, use a map of the United States to trace the geologic periods of time. Imagine traveling out of the past on a "time train" from New York to San Francisco. As you leave New York, the first dinosaurs and flying reptiles (Triassic period) have already developed. Somewhere between Ohio and Nebraska, you look out your window and see huge plant-eating dinosaurs as big as buses and meat-eating dinosaurs walking on their hind legs (Jurassic period).

Between Nebraska and Utah (Cretaceous period) you might see Tyrannosaurus rex stalking other dinosaurs. As you cross Nevada, you notice that the dinosaurs have mysteriously disappeared and that birds and mammals have taken over. Somewhere near the California border, you see a few of the first humans and watch them develop. About halfway across the state, you notice that humans began keeping animals and growing crops. In San Francisco, you see things as they are today. You can see that dinosaurs were around many times longer than humans have been on the earth.

Look up dinosaurs in an encyclopedia. Do you think dinosaurs disappeared through sudden or gradual extinction? Could they have been cold-blooded, warm-blooded, or both? Some scientist think that birds developed from small meat-eating dinosaurs.

did you ☆ know?

❏ That dinosaurs were the largest creatures to ever walk the earth. Some dinosaurs were nearly half the length of a football field. Others were smaller than a chicken.

❏ That fossilized remains of dinosaurs have been found on every continent in the world.

❏ That *sauropod* means "lizard feet" and that sauropods had five toes. However, a sauropod's foot would have looked more like an elephant's foot than a lizard's.

Part 6
The young physicist

Physics is the study of the natural world around us. It tells us how and why a lever can lift a heavy weight, why hot air rises, and what light is. It is the study of electricity, magnetism, and how sound waves travel. Physics tells us how and why a mirror works, why we can't see in the dark, and how birds are able to fly. This fascinating science covers such a wide range of subjects that no simple definition explains it.

Prehistoric people first applied the laws of physics when they put rollers under heavy loads and discovered that the load was easier to pull. Early Greek philosophers, the great thinkers of their time, proposed that Earth was a sphere, and guessed that an eclipse occurred when Earth blocked the Sun's rays.

Around 400 B.C., one philosopher taught that matter was made up of tiny particles he called *atoms*. During the second century B.C., the laws of the lever were discovered along with the principles of the weight of floating bodies. During the Middle Ages, however, few people were interested in physics, and little progress was made for hundreds of years. Fortunately, in the 1200s and 1400s, people such as Roger Bacon and Leonardo da Vinci began to see the importance of physics and, by the mid-1500s, the science of physics was born.

31
When was the earth first described as a huge magnet?

materials ☆
- ❏ Magnetic compass
- ❏ Paper
- ❏ Bar magnet
- ❏ Pencil

procedure ☆   1. Use a compass to draw a circle on a piece of paper with the diameter a little larger than the length of the bar magnet. The circle represents the earth.

Draw a circle to represent the earth.

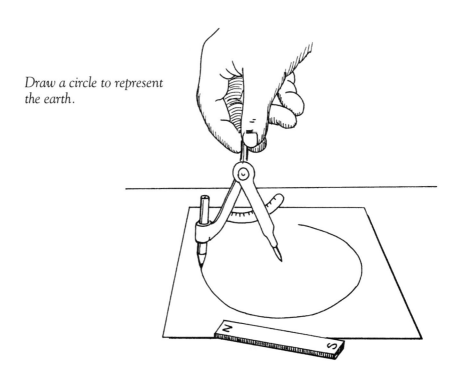

2. With a pencil, make a mark at the top and bottom to represent the North Pole and South Pole.

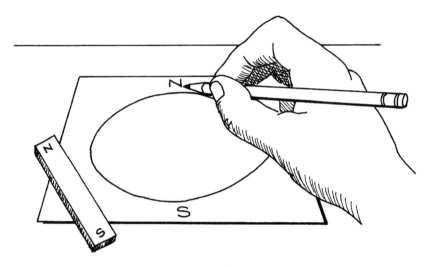

Mark the earth's poles.

3. Place the bar magnet in the circle and align the north and south poles of the magnet with the North and South Poles you marked on the earth.

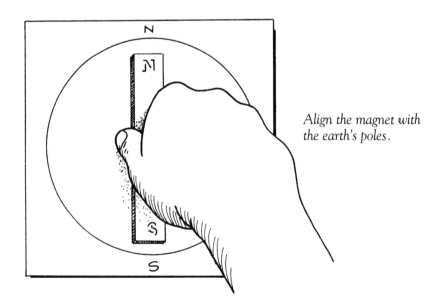

Align the magnet with the earth's poles.

4. Place the magnetic compass outside the circle near the South Pole. Slowly move the compass in the direction indicated by its needle. Continue moving the compass until it gets to the North Pole.

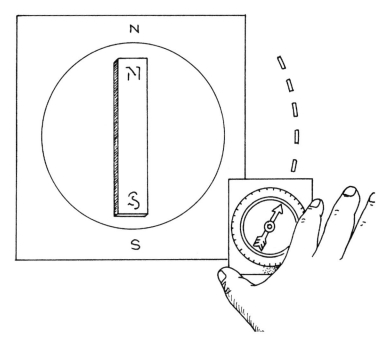

Move the compass toward the North Pole.

results ☆ The compass travels in a circular path from the South Pole to the North Pole. This path is the pattern of the earth's magnetic field. The earth's magnetic axis is not the same as the true north-south axis. The true North Pole is located at 90 degrees north latitude and all of the degrees north longitude.

The magnetic north pole is at 76 degrees north latitude and 100 degrees west longitude. It is located near Prince of Wales Island in Canada about 1000 miles (1609 km) from the true North Pole. The magnetic pole is not on the surface but is buried about 70 miles (113 km) inside the earth. The magnetic south pole is at 66 degrees south latitude and 139 degrees east longitude. It is about 1500 miles (2414 km) from the true South Pole, directly south of Sydney, Australia.

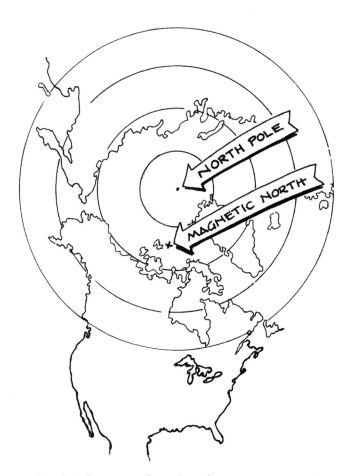

The magnetic North Pole is in northern Canada.

William Gilbert, a physician to England's Queen Elizabeth I, first began to use scientific methods to study magnetic phenomena. In 1600, Gilbert published his discovery that the earth itself is a huge magnet.

further ☆ studies
Look at a map of the world and find the location of the magnetic north pole. Does the magnetic north pole have anything to do with the mystic display of colors called the northern lights? Do birds have the ability to use the earth's magnetic field for navigation? How do whales find their way?

did you ☆
know?

❏ That at the magnetic poles the lines of force are vertical, but near the equator they are horizontal.

❏ That the magnetic poles drift somewhat and navigational charts must be upgraded every few years.

❏ That in the history of the earth, its magnetic field has had many complete reversals of the poles.

32

When was the galvanometer discovered?

materials ☆
- ❏ A 3-×-4 inch (7.6-×-10.0 cm) piece of cardboard
- ❏ 18-gauge wire (bell wire)
- ❏ Magnetic compass
- ❏ Flashlight battery
- ❏ Wire strippers

procedure ☆ 1. Ask an adult to help you with the following procedures. Fold two ends of the cardboard up to form supports for the wire.

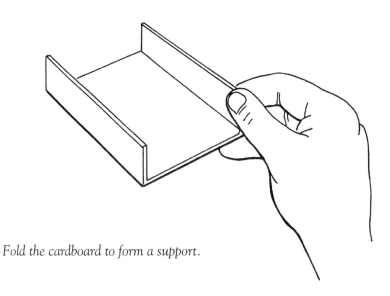

Fold the cardboard to form a support.

2. Wrap the wire around the cardboard about 30 times and leave about a foot (30.5 cm) of surplus wire for connections. Ask an adult to strip about ½ inch (1.27 cm) of insulation from the ends of the wires.
3. Place the compass on the cardboard and beneath the wires. Then turn the cardboard so that the wires run east and west.

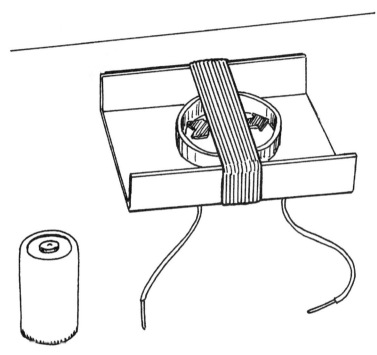

Align the wires east and west.

4. Touch the ends of the wires to the battery and observe the compass needle.

results ☆ The compass needle should be deflected nearly east and west. The needle movement indicates a current flow in the wires. A *galvanometer* is a meter that can detect and measure the strength of an electric current. The stronger the current, the further the needle movement. People once called electrical current *electricity galvanism* in honor of Luigi Galvani. Galvani experimented with current electricity in 1786.

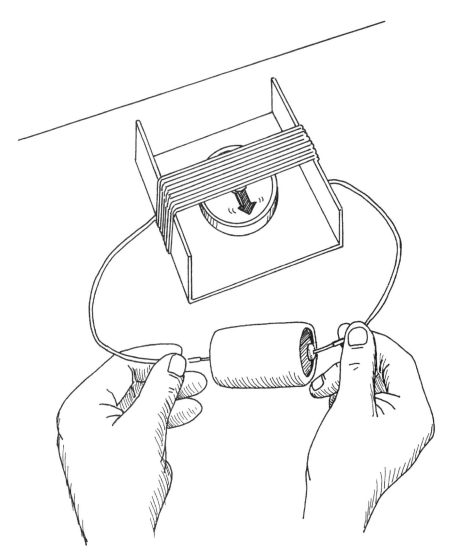

Touch the wires to the battery.

The principle of the galvanometer was first discovered in the winter between the years 1819 and 1820. Hans Christian Oersted, a physics professor at the University of Copenhagen, was conducting a class to show that an electrical current could produce heat in a wire (a principle used by our toasters today). By design or by chance, a magnetic compass had been left near the wire. When the current was applied to the wire to produce heat, the compass needle suddenly pointed directly to the wire. When the battery was disconnected, the needle swung back to the north-south direction. Oersted had discovered the principle, called *electromagnetism*, that allows the galvanometer to work.

further ☆
studies
Look up *electricity* and *motor* in an encyclopedia. Can an electric current flow without producing a magnetic field? Can a magnetic field be used to produce electricity? Wrap several turns of a wire around the compass. Wrap several turns of another wire into a coil. Connect the two wires together and slide a bar magnet in and out of the open coil. Does the needle move? Have you made a *generator*?

did you ☆
know?
❒ That the static on a radio and the brief lines across a TV screen during a lightning flash are caused by electromagnetic waves generated by the current of the lightning.
❒ That if a car radio is tuned to an AM station, and you pass under power lines, you hear the static of the magnetic field of the wires overhead.

33
When was the
first battery made?

materials ☆
- ❏ Lemon
- ❏ Knife
- ❏ Table
- ❏ Iron or steel wire
- ❏ Copper wire
- ❏ Tablespoon
- ❏ Lemon juice or vinegar
- ❏ Glass
- ❏ Water
- ❏ Paper towel
- ❏ Ten dimes and ten pennies
- ❏ Galvanometer from the preceding activity or a dc voltmeter

procedure ☆

1. Ask an adult to help you with the following procedures. Roll the lemon on a table to make the lemon juicy. Then cut the lemon in half.
2. Insert the iron wire in one side and the copper wire in the other. Connect the wires to the galvanometer.
3. Mix a tablespoon of lemon juice or vinegar with a glass of water. Soak the paper towel in the solution and then tear it into coin-size pieces.
4. Place a penny on a table then lay a wet piece of paper towel on top. Place a dime on top followed by another wet paper. Stack alternating pennies and dimes (with each coin separated by wet paper). The stack should start with one coin (the penny on the bottom) and end with the other (a dime) on top. Connect the stack to the wires of a galvanometer.

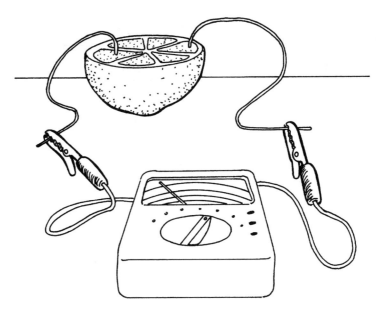

A lemon battery

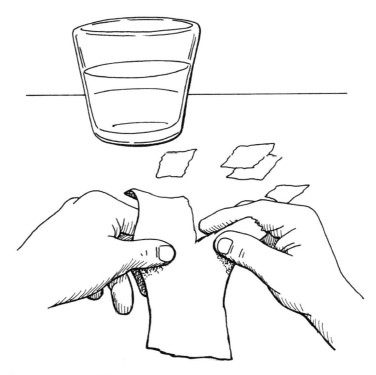

Tear the paper into small pieces.

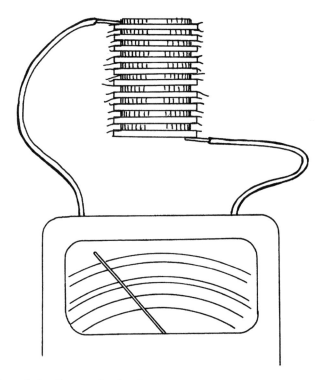

The voltaic pile was the first battery.

results ☆ In each case, the galvanometer (or *voltmeter*) should indicate a small amount of voltage. In 1780, Luigi Galvani conducted a class in anatomy at the University of Bologna. He dissected a frog next to a generating machine that had been used in an earlier experiment. Suddenly, a spark passed between the frog and the machine. Galvani thought he had discovered an electrical source in animals.

Alessandro Volta, a professor of physics at the University of Padua, disagreed with Galvani. He believed electricity came from the difference between two metals. In 1800, Volta constructed a stack of silver and zinc disks separated by leather or paper disks soaked in some kind of solution such as salt water or perhaps vinegar or lemon juice. This stack was know as the *voltaic pile* and is considered to have been the first battery.

Italian physiologist and physicist Luigi Galvani

Italian physicist Alessandro Volta

There was some truth in both men's convictions. The chemicals in the frog's leg aided in the generation of the spark, but you need the two different kinds of metals as well as the chemicals.

further studies ☆ Can you think of other sources of electricity? Substitute a potato for the lemon. Can a potato produce electricity? Does a battery produce an electrical current by chemical action? Would smaller, more powerful batteries help in developing a practical electric car?

did you know? ☆ ❐ That in the United States, Thomas Davenport built a working model of an electric car in 1836. And by 1900, electric cars were the most common type of automobile. They were popular because they had silent motors and did not give off poisonous fumes. They lost out to gasoline engines because the batteries had to be recharged about every 100 miles (160 km).

❐ That batteries today can weigh less than 0.02 of an ounce.

❐ That some batteries must be disposed of as hazardous waste because of the chemicals inside.

34
When does light bend?

materials ☆
- ❏ Shallow bowl
- ❏ Table
- ❏ Coin (nickel or quarter)
- ❏ Glass of water

procedure ☆
1. Place the bowl on the table near one edge. Put the coin in the bowl, toward you, at the edge where the bowl starts to slope upward. This spot is a reference point.

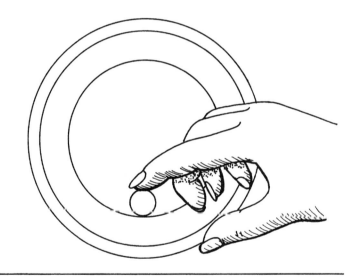

Place the coin at the inside edge of the bowl.

2. Look at the coin from an angle above the lip of the bowl. Position your head so that you can only see the outer edge of the coin.
3. Keep looking at the coin and slowly pour water into the bowl.

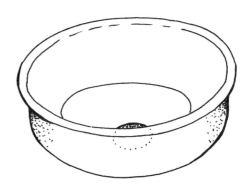

Look at the tip of the coin.

Add water to the bowl.

results ☆ The coin appears to move toward the center of the bowl until it is in full view. When no water is in the bowl, the light travels in a straight line from the coin to your eyes. As you sight over the lip of the bowl, only the outer edge of the coin reflects light to your eyes. But when water is added, the light rays are bent, or refracted, by the water. The angle of bending is great enough to allow all of the light reflected from the coin to travel to your eyes.

further ☆
studies

Place a pencil in a glass of water. Allow it to slant across the inside of the glass. Does the pencil appear to be broken where it enters the water? Does it seem larger below the water? Are the light rays from below the water bent before they reach your eyes?

Compare the differences between round jars and square jars. Fill both jars about half full of water and lower a pencil into each jar. Move each pencil toward and away from you. Is there a difference between the jars? Does the round jar magnify the pencil?

Place the pencil in a glass of water.

did you ☆
know?

❑ That food companies often pack items such as cherries and olives in round jars to make the food appear larger.
❑ That the speed of light is the fastest possible speed: 186,282 miles (299,792 km) per second. If the North Star went out, we would not know it for about 460 years.

35
When can
water magnify?

materials ☆
- ❏ Pencil
- ❏ Piece of fine wire about 6 inches long
- ❏ Bowl of water
- ❏ Printed page

procedure ☆
1. Use the pointed end of a pencil to bend a round loop about ¼ inch (6.4 mm) in diameter in one end of the wire. Twist the wires together to form a closed loop.
2. Dip the wire loop in the water and carefully remove it.
3. Look through the loop at a printed page.

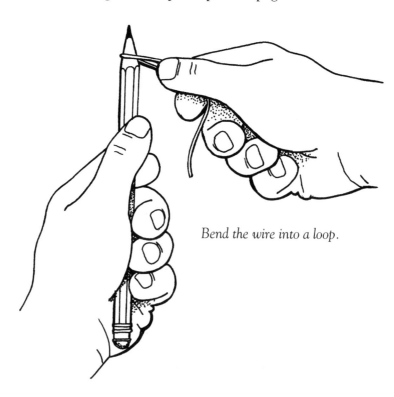

Bend the wire into a loop.

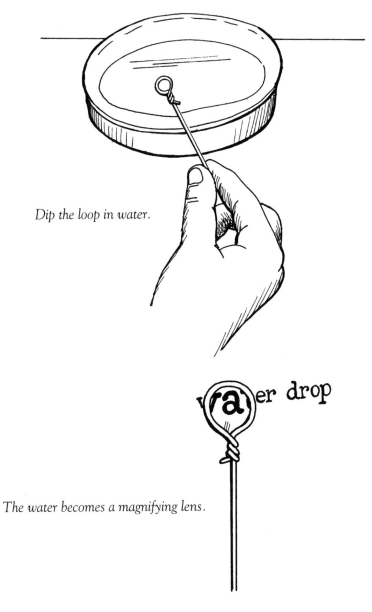

Dip the loop in water.

The water becomes a magnifying lens.

results ☆ Some water should stay in the loop when it is removed from the bowl of water. The circle of water acts like a magnifying lens. Most lenses are made of glass, but any transparent material can be used. In your experiment, both sides of the water in the loop curve outward and form a *convex lens*. Convex lenses are used in magnifying glasses.

CONVEX LENS

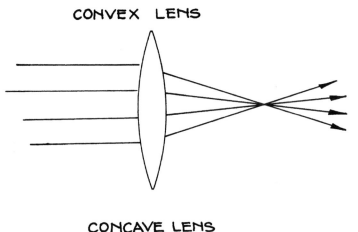

CONCAVE LENS

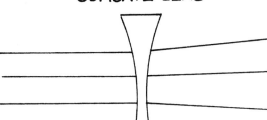

Two types of lenses

further studies ☆ If a convex lens is thicker in the middle, what is a *concave lens*? Does a convex lens converge light rays to a focusing point? Do light rays spread after passing through a concave lens? Why does water in a swimming pool always looks shallower than it is?

did you know? ☆ ❏ That lenses are specially shaped pieces of glass that refract, or bend, light in a precise way.
❏ That nineteenth century physicist James Clerk Maxwell showed that light was only one form of electromagnetic radiation. There are other kinds of radiation, from radio waves to cosmic rays, but light is the only type of radiation we can see.

Scottish physicist James Clerk Maxwell

36
When can sunlight distill water?

materials
- ❐ Sunny day
- ❐ Shovel
- ❐ Bowl
- ❐ Sheet of clear plastic
- ❐ Marbles or small stones

procedure ☆

1. Ask an adult to help you with the following procedures. On a sunny day, dig a hole in the ground about 12 inches (30.5 cm) deep and about 18 inches (46 cm) across.
2. Place the bowl in the center of the bottom of the hole.
3. Spread the plastic over the hole. Place a marble or stone in the center of the plastic and allow the plastic to sag into the hole slightly above the bowl.
4. Anchor the plastic by placing stones around the edge. Cover the edges of the plastic with dirt to trap the air in the hole.

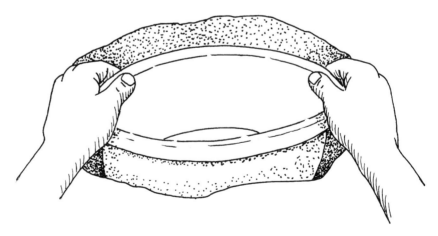

Place the bowl in the hole.

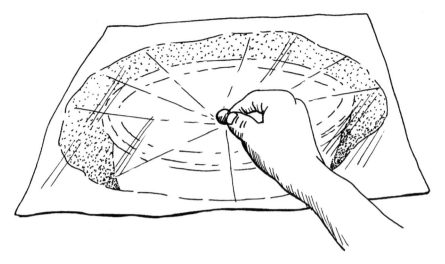

Place the marble on the plastic.

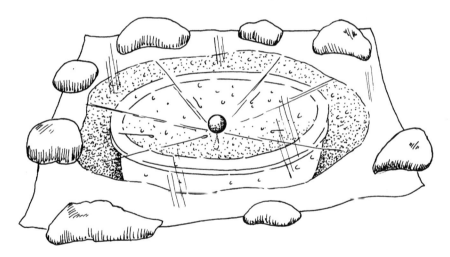

Anchor the plastic with stones.

results ☆ After several hours, depending on the moisture in the ground and the heat of the sun, water begins to drip into the bowl. Sunlight shines through the plastic and warms the earth inside the hole, causing the moisture in the ground to evaporate. The water vapor then condenses and forms droplets on the underside of the plastic. The drops then run down to the lowest point and drip into the bowl.

Distill means to fall in drops or to purify or refine something through distillation. *Distillation* is a process in which a liquid is first heated and then cooled to allow it to condense.

further ☆ studies How can sea water be used for drinking water? Why is distilled water purer than ground water? Is rain or dew a form of water that has been distilled by nature?

did you ☆ know?
❏ That light and heat from the Sun can be converted to *solar energy*. The Romans gave the Latin name *Sol* to the Sun.
❏ That coal and oil are the results of trapped sunlight. Just like today, millions of years ago plants lived and trapped the Sun's energy. When the plants died, some sank down in swampy soil and slowly over time and under great pressure changed into coal. Today, when we burn coal or drive down a highway, we are using the energy from the Sun that the plants trapped millions of years ago.

37
When does sound tell the direction of a moving object?

materials ☆ ❑ An approaching train, fire truck, or ambulance

procedure ☆ 1. Listen to the sound of a train or an emergency vehicle. Notice the difference in the pitch of the sound as it passes. Now compare the pitch of the sound as it moves away.

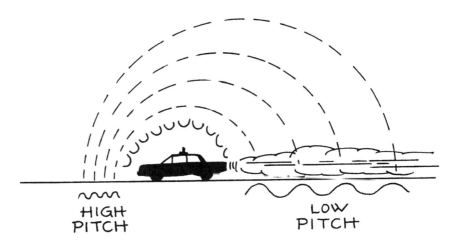

HIGH PITCH

LOW PITCH

Sound waves from something approaching are pushed together. Sound waves from something moving away are stretched apart.

results ☆ The *pitch* or *frequency* of a train seems to sound higher as it comes closer, and then becomes lower as it goes away. This change happens because, as the train approaches, the sound waves are pushed closer together. More sound waves strike your ear each second. The frequency is higher; therefore, the pitch seems higher. When the train passes and starts moving away, the sound waves are stretched apart. Fewer sound waves strike your ear each second. The frequency is lower and the pitch sounds lower.

If you were able to move with the train, you would know that it only has one pitch. The change in pitch is called the *Doppler effect*, after the German physicist Christian Doppler. Doppler described the principle in 1842. The Doppler effect is the apparent change in frequency of sound, light, or radio waves caused by motion.

did you ☆ know?
- ❐ That every sound you hear is created by something vibrating.
- ❐ That the high-pitched squeaks bats use like radar is an *ultrasonic* or high frequency, and is far above the frequency that humans can hear.

38
When does a drinking glass sing?

materials ☆   ❑ Water
          ❑ Thin drinking glass
          ❑ Vinegar

procedure ☆   1. Pour a little water into the glass and add a few drops of vinegar.
            2. Dip one of your fingers into the water and slowly rub it around the rim of the glass. Use a smooth, steady motion. You might have to dip your finger back into the water a few times.

Slowly rub your finger around the rim of the glass.

results ☆ Within two or three tries, you should produce musical sounds. The vinegar is used to remove the oil from the finger tip and the rim of the glass. The moving finger produces vibrations in the glass the same way a violin bow produces vibrations on the strings of a violin.

further studies ☆ Try using different sizes of glasses and different amounts of water. Does the pitch change? Get a friend to help you make an identical sound on another glass. Does the sound become louder or fade away at intervals?

Did you know? ☆
- ❒ That *sonar*, used by ships and submarines, is made up from the words "sound navigation and ranging." *Radar* comes from "radio detection and ranging." The units send out sound waves and measure any echoes that return. Sonars and radars are echo-ranging devices.
- ❒ That if you pluck the string of a guitar and another guitar is close enough, the same note vibrates on the second guitar.
- ❒ That the human brain can measure the differences in the speed of sound. Ask someone to stand first behind you and then on one side as they make a small sound. You can tell which shoulder they are behind because of the differences it takes for the sound to reach each ear.

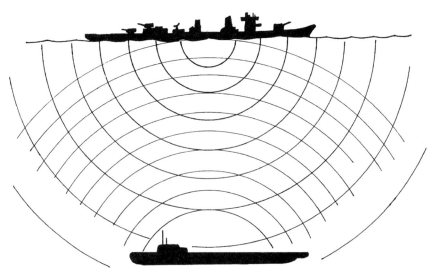

Sonar is used by ships and submarines.

Index

Illustration numbers are in **boldface**.

About the Author

Robert W. Wood has professional experience in aviation science, experimental agriculture, electricity, electronics, and science research. He is the author of more than a dozen physics and science books, as well as several home maintenance books. His work has been featured in major newspapers and magazines, and he has been a guest on radio talk shows around the United States. Some of his books have been translated into other languages, including Turkish.

Bob enjoys the wonders of nature and is always interested in the advances of science and how it affects our lifestyles. Although he has traveled and worked worldwide, he now lives with his wife and family in Arkansas.